PLUMBING AND DOMESTIC HEATING

Other titles from E & FN Spon

SPON'S

CONTRACTORS' HANDBOOK

PLUMBING AND DOMESTIC HEATING

1995

Tweeds

CHARTERED QUANTITY SURVEYORS,
COST ENGINEERS, CONSTRUCTION ECONOMISTS

E & FN SPON

An Imprint of Chapman & Hall

London · Glasgow · Weinheim · New York · Tokyo · Melbourne · Madras

Published by E & FN Spon, an imprint of Chapman & Hall, 2–6 Boundary Row, London SE1 8HN

Chapman & Hall, 2–6 Boundary Row, London SE1 8HN, UK

Blackie Academic & Professional, Wester Cleddens Road, Bishopbriggs, Glasgow G64 2NZ, UK

Chapman & Hall GmbH, Pappelallee 3, 69469 Weinheim, Germany

Chapman & Hall USA, One Penn Plaza, 41st Floor, New York NY 10119, USA

Chapman & Hall Japan, ITP-Japan, Kyowa Building, 3F, 2-2-1 Hirakawacho, Chiyoda-ku, Tokyo 102, Japan

Chapman & Hall Australia, Thomas Nelson Australia, 102 Dodds Street, South Melbourne, Victoria 3205, Australia

Chapman & Hall India, R. Seshadri, 32 Second Main Road, CIT East, Madras 600 035, India

First edition 1986
Sixth edition 1994

Printed in Great Britain by St Edmundsbury Press
Bury St Edmunds, Suffolk

ISBN 0 419 18580 1

A catalogue record for this book is available from the British Library

Printed on permanent acid-free text paper, manufactured in accordance with ANSI/NISO Z39.48-1992 and ANSI/NISO Z39.48-1984 (Permanence of Paper).

CONTENTS

Preface

This is the sixth edition of Spon's Plumbing and Domestic Heating Contractors' Handbook: the book is intended to help the plumber or heating engineer who is on the point of becoming self-employed or has already taken the plunge and is now operating a small contracting business.

Well established plumbing contractors can also benefit by using this book and, from the helpful and constructive criticisms received, it would appear they are among the most regular users of the wealth of information contained in the following pages.

Advice on the business side of the industry is contained in the first three chapters and is set in an easy-to-read style with many examples. The information given on current tax rates is taken from the spring budget proposals. Readers should check whether any changes were made to the proposals during the passage of the Finance Bill through Parliament.

It is not intended that these chapters should replace the need for professional advice when the occasion warrants it, but they are meant to complement this need and hopefully save money in consultation fees.

Chapter 4 sets out some of the basic principles of estimating but the main thrust of the book lies in Chapter 5 - Rates for measured work. The information contained in this chapter is intended as a base on which a plumbing estimator can produce his quotations.

It cannot be stressed too strongly that, despite claims made on their behalf, price books should not be used as a literal source of information for preparing quotations, but as a base on which a contractor can formulate his own unique pricing data.

The measured items are presented in accordance with the requirements of the 7th Edition of the Standard Method of Measurement of Building Works (SMM7). Although this may be irrelevant to the needs of smaller contractors it may well be required by those larger firms tendering for work on tender documents based upon SMM7.

PREFACE

The indexing and reference systems have been improved so that the self-employed plumber who prices jobs without formal tender documents can also receive the maximum benefit from the book.

It should be noted however that not all of the requirements of SMM 7 have been observed and it is hoped that the compromise will produce a balance which will suit the majority of readers.

There are many items affecting the value of plumbing works. The speed and skill of individual workmen and the price paid for materials are factors which can directly affect the profitability of a job. The rates in Chapter 5 should be used as a basis only for preparing an estimate or quotation.

The construction industry is going through a particularly difficult period and many contractors are submitting tenders which have almost been reduced to cost with the minimum allowance for profit and overheads. This policy is damaging to both the contractor and the client and is deplored.

The pricing levels in this series of books has remained constant throughout the recession so that regular users are provided with a constant data base of cost information on which to make their bids. In general, the material prices are 'trade prices'. It is acknowledged that substantial discounts can be obtained from merchants and suppliers by long-standing customers but these can vary so widely that I believe it would be misleading to attempt to incorporate them into this book.

These comments do not detract from the real value of the information Chapter 5. Over 4,000 item descriptions and unit rates are given which are based on 25,000 separate pieces of data providing a wealth of detailed cost information.

There are now many women working in the construction industry; the pronoun 'he' used throughout this book applies to both men and women.

The editors have received a great deal of help from many sources in the research and preparation of this book and would like to acknowledge the assistance given with grateful thanks.

Drainage Systems
IMI Yorkshire Fittings Ltd
Fulbora Ltd
'Osma' Wavin Building Products
Marley Extrusions Ltd
Key Terrain Ltd
BMMR
Potterton International Ltd
Jewsons
Alumasc Ltd
Stelrad Group Ltd
Myson Ltd
Mira-Walker Crosswell
Dorothy Spain
Paul Spain
Caradon Twyfords

PREFACE

Grant Thornton provided the information and advice on business and taxation matters set out in Chapters 1 to 3.

The editors would welcome constructive criticism of the book together with suggestions for improving its scope and contents. Whilst every effort has been made to ensure the accuracy of the information given in this publication, neither the editors nor the publishers in any way accept liability of any kind resulting from the use made by any person of such information.

Bryan J.D. Spain FInstCes, MACostE
TWEEDS (incorporating SPAIN AND PARTNERS)
Cavern Walks
8 Matthew Street
Liverpool L2 6RE

The Institute of Plumbing

64 Station Lane Hornchurch Essex RM12 6NB
Tel: Hornchurch (0708) 472791 Fax: (0708) 448987

I am delighted to have been given the opportunity to produce the foreword to the 6th edition of *Spon's Contractors' Handbook Plumbing and Domestic Heating*.

As we move out of one of the worst recessions of our time it is more important than ever that those involved with estimating and tendering have accurate cost data.

The business advice contained within the 6th edition will prove to be of particular benefit to those about to venture into business.

I have pleasure in commending this book to you and am sure you will not regret your purchase.

Richard A. Soper

Richard A. Soper
President – The Institute of Plumbing

SMM6/SMM7 Table

SMM6	SMM7
	H CLADDING/COVERING - SMALL WORKS
	N FURNITURE AND EQUIPMENT
R PLUMBING AND MECHANICAL ENGINEERING INSTALLATIONS	R DISPOSAL SYSTEM ENGINEERING INSTALLATIONS
	S PIPED SUPPLY SYSTEMS
	T MECHANICAL HEATING/COOLING REFRIGERATION SYSTEMS
	U VENTILATION/AIR CONDITIONING SYSTEMS
	X TRANSPORT SYSTEMS
	Y MECHANICAL AND ELECTRICAL SERVICES MEASUREMENT
Classification of work	
Sheet metal flashings and gutters	H71 Lead sheet coverings/flashings
Equipment and ancillaries	N13 Sanitary appliances/fittings

SMM6/SMM7 TABLE

SMM6	SMM7
a. rainwater installation	R10 Rainwater pipework/gutters
b. sanitary installation (including traps)	R11 Foul drainage above ground
c. cold water installation	S10 Cold water
	S13 Pressurized water
	S14 Irrigation
	S15 Fountains/Water features
	S20 Treated/Deionized/Distilled water
	S21 Swimming pool water treatment
d. firefighting installation	S60 Fire hose reels
	S61 Dry risers
	S62 Wet risers
	S63 Sprinklers
	S64 Deluge
	S65 Fire hydrants
	S70 Gas fire fighting
	S71 Foam fire fighting
e. heated, hot water installations etc.	S11 Hot water
	S12 Hot and cold water (small scale)
	S51 Steam
	T20 Primary heat distribution
	T30 Medium-temperature hot water heating
	T31 Low-temperature hot water heating
	T32 Low-temperature hot water heating (small scale)
	T33 Steam heating
f. fuel oil installation	S40 Petrol/Oil lubrication
	S41 Fuel oil storage/distribution

SMM6/SMM7 INDEX

SMM6	SMM7
	S32 Natural gas
g. fuel gas installation	S33 Liquid petroleum gas
h. refrigeration installation	T61 Primary/secondary cooling distribution T70 Local cooling units T71 Cold rooms T72 Ice pads
j. compressed air installation	S30 Compressed air S31 Instrument air
k. hydraulic installation	-
l. chemical installation	-
m. special gas installation	S34 Medical/Laboratory gas
n. medical suction installation	S50 Vacuum
p. pneumatic tube installation	S50 Vacuum
q. vacuum installation	R30 Centralized vacuum cleaning S50 Vacuum
r. refuse disposal installation	R14 Laboratory/Industrial waste drainage R20 Sewage pumping R21 Sewage treatment/sterilization R31 Refuse chutes R32 Compactors/Macerators R33 Incineration plant
s. air-handling installation	T40 Warm-air heating T41 Warm-air heating (small scale) T42 Local heating units T50 Heat recovery U10 General supply/extract U11 Toilet extract U12 Kitchen extract U13 Car-parking extract U14 Smoke extract/Smoke control

SMM6/SMM7 INDEX

SMM6	SMM7
	U15 Safety cabinet/Fume cupboard extract
	U16 Fume extract
	U17 Anaesthetic gas extract
	U20 Dust collection
	U30 Low-velocity air conditioning
	U31 VAV air conditioning
	U32 Dual-duct air conditioning
	U33 Multi-zone air conditioning
	U40 Induction air conditioning
	U41 Fan-coil air conditioning
	U42 Terminal re-heat air conditioning
	U43 Terminal heat pump air conditioning
	U50 Hybrid-system air conditioning
	U60 Free-standing air conditioning units
	U61 Windows/Wall air conditioning units
	U70 Air curtains
	Y24 Trace heating
	Y25 Cleaning and chemical treatment
t. automatic control installation	Y53 Control components - mechanical
u. special equipment e.g. kitchen equipment	N12 Catering equipment
v. other specialist installations	X10 Lifts
	X11 Escalators
	X12 Moving pavements
	X20 Hoists
	X21 Cranes
	X22 Travelling cradles
	X23 Goods distribution/Mechanized warehousing

SMM6/SMM7 INDEX

SMM6	SMM7
	X30 Mechanical document conveying
	X31 Pneumatic document conveying
	X32 Automatic document filing and retrieval
Gutterwork	R10 Rainwater pipework/gutters
Pipework (including fittings)	Y10 Pipelines
Ductwork	Y30 Air ductlines
	Y31 Air ductline ancillaries
	Y40 Air-handling units
	Y41 Fans
	Y42 Air filtration
	Y43 Heating/Cooling coils
	Y44 Humidifiers
	Y45 Silencers/Acoustic treatment
	Y46 Grilles/Diffusers/Louvres
Equipment and ancillaries	N13 Sanitary appliances/fittings
	T10 Gas/Oil-fired boilers
	T11 Coal-fired boilers
	T12 Electrode/Direct electric boilers
	T13 Packaged steam generators
	T14 Heat pumps
	T15 Solar collectors
	T16 Alternative fuel boilers
	T60 Central refrigeration plant
	Y11 Pipeline ancillaries
	Y20 Pumps
	Y21 Water tanks/cisterns
	Y22 Heat exchangers
	Y23 Storage cylinders/calorifiers
	Y52 Vibration isolation mountings
	Y54 Identification - mechanical
Insulation	Y50 Thermal insulation

SMM6/SMM7 INDEX

SMM6	SMM7
Sundries	Y51 Testing and commissioning of mechanical services Y59 Sundry common mechanical items
Builder's work	P31 Holes/Chases/Covers/Supports for services
Protection	A42 Contractor's services and facilities

Introduction

There are two main ways that contractors prepare tenders and quotations. The first and probably the less frequent method is to insert rates against item descriptions in a tender document prepared by the prospective employer's professional advisers. This would occur on major schemes where a bill of quantities and/or a schedule of rates has been prepared. In this case the contractor would be able to examine and use the rates contained in Chapter 5 of this book as a basis for his bid. Usually, however, the contractor is either handed a plan and specification or merely invited to inspect the premises and prepare his offer without the benefit of any paperwork at all! In either of these two cases the contractor must take off his own quantities. Once he has done that, the contents of Chapter 5 can be used as a base on which he can prepare his quotation.

Whichever method is used the main value of this book lies in a sensible application of the thousands of rates in Chapter 5. People who have not used price books before sometimes query their value in that every craftsman does not work at the same speed, wide variations in discounts for materials can be obtained, each job has different cost related circumstances etc.

The answer to these criticisms is that successful users of price books are fully aware of these difficulties but overcome them by understanding the relationship between their own production costs and material discounts and those assumed in the book.

For example, a regular user of this book will know that the published rates may be a certain percentage higher or lower than his own costs. With this knowledge he can quickly prepare his quotations by using the book rates and making the appropriate percentage adjustment at the end.

The editors regularly receive comments from contractors using these books and it seems that most firms' costs fall in a band between 10% higher and lower than those in this book. Some firms have been good enough to state that they have been able to win contracts in competition by using the book's rates verbatim and have secured good profits as well.

Careful thought must be given to the unique circumstances of each job undertaken, and the percentage allowances for 'project labour factors' at the beginning of Chapter 5 should be studied carefully and their potential cost in all quotations.

Profits can turn into losses without an assessment of the effect that each project's elements can have on the rates and the editors urge readers to consider this matter very carefully.

Someone once said that price books were like guns - dangerous in the wrong hands! There is some truth in this but using the information wisely can save the one commodity you cannot buy - time.

And that in itself should be of great value to a busy contractor.

Chapter 1
Starting up in Business

Before committing himself by giving up his job, the would-be businessman should consider carefully whether he has the skills and also the temperament to survive in the highly competitive self-employed market. He should also do a lot of research and seek out as much information as possible about how to run a business. He should then know whether his business idea is likely to work in practice and have some idea of the new, strange and sometimes complex requirements of running such a business.

INITIAL RESEARCH

Matters to be researched should include the following.

Finance

Assess what funds will be needed and when will items such as premises, plant, transport, tools, initial stocks of materials, wages, overheads and the proprietors' living expenses need to be paid for before the cash from work done begins to flow in. How is the price of work done to be calculated?

All of this needs a proper 'business plan' and if the bank or other sources of finance are to be tapped such a plan is essential. Fortunately help in its preparation is available from a number of agencies sponsored by goverment, local authorities and industry, usually at little or no cost.

Testing the market

Talk to as many traders already operating in the same field as possible. Try to identify if the need is in the industrial, commercial, local government or domestic field. Talk to likely customers and clients and consider whether it is possible to improve on what they are being offered at present in terms of price, quality, speed, convenience, follow-up, etc.

Advertising

Those entering the domestic side of the business will need to think about the best way to reach potential customers. Are local word-of-mouth recommendations enough to provide reasonably continuous work? If not, what would be the most effective method?

Advertising is costly and it is a waste of funds to place an advert in a paper which circulates in areas A, B, C and D if the business can only cover area A. Advice on the best medium and the content of adverts is available at the Small Business Service.

Experience and training

After making an objective appraisal of the likely market for the goods or services, then the extent to which one is equipped to satisfy its requirements should be considered. Gaps in experience might be filled by a change of present duties or of employer, whilst some lack of skills can probably be overcome by taking a training course.

INITIAL INFORMATION

Training and Enterprise Councils (TECs)

There is no shortage of information about the many aspects of starting and running your own business: finance; marketing; legal requirements; developing your business idea; taxation, etc, are all the subject of a mountain of books, pamplets, guides and courses. Indeed the likelihood is that the aspiring businessman will be overloaded and thoroughly confused rather than left high and dry without guidance. Nor is it necessary to pay out a lot of money.

A good place to start for both information and advice is your local TEC. This comprises a board of directors drawn from the top men in local industry, commerce, education, trade unions etc, who, together with their staff and experienced business counsellors, assist both new and established concerns in all aspects of running a business. This takes the form not only of across-the-table advice but also, if desired, hands-on assistance in management, marketing, finance etc. There are also training courses and seminars available in most areas.

Contact can be made through the local job centre, Citizens Advice Bureau or by ringing Freephone Enterprise 0800 222999.

INITIAL INFORMATION

Business links

There are organisations currently being established with a view to providing a 'one-stop-shop' for advice and assistance to owner-managed businesses. When established they will often replace the need to contact TECs and many of the other offical organizations listed below.

Enterprise Agencies

Another useful source of free help and advice is the local Enterprise Agency. Run by local businesses for small and developing concerns it covers similar ground to the TEC.

The address and telephone number of the agency in your area can be obtained by ringing 'Business in the Community' on 071-253 3716.

The Business Start-up scheme

This is an allowance of £40 a week, in addition to any income made from your business, paid for forty weeks.

To qualify you must be at least eighteen and under sixty-five, work at least thirty six hours per week in the business and have been unemployed for at least six weeks or fall into one of the other categories - disabled, ex-HMF, redundant etc.

The first step is to get the booklet on the subject from your local job centre or TEC; all the details are in it, including how and where to apply. Once in receipt of the Enterprise Allowance you will also have the benefit of advice and assistance from an experienced businessman from your TEC. All the initial counselling services and training courses are free.

Potential customers and trade contacts

Many who become self-employed in the construction industry already have experience as employees. Use these contacts to check the market, establish the sort of work which is available and the current contract rates. In the domestic market check on the competition for prices, standards of work and service provided, customer complaints and types of advertising.

Try to get firm promises of work before the start-up date.

Potential suppliers

Canvass local suppliers for the best prices, credit terms, minimum order, discounts offered and delivery times. Remember cash is the life blood of business and a supplier who gives 30 days' credit and delivers small quantities at 24/48 hours' notice may be a better buy than one with lower prices but who operates on cash and carry terms only. It is important not to overstock, but carry only what is needed for current requirements. Do not tie up scarce and expensive money in stock which may not be used for weeks or even months.

Banks

Approach banks for information about a business current account and financial services available. Find out about the types of loan required; from an overdraft for working capital to medium and long terms loans including the Government's Loan Guarantee Scheme (see page 10) for the purchase of plant and machinery and alterations to premises. See the TEC first for information about the best approach and find out what the Bank Manager will need. Shop around several banks and branches if you are not satisfied at first; managers vary widely in their views of what is a viable business proposition. Most banks have useful free information packs to help business start-up.

Point of contact: the local bank manager.

HM Inspector of Taxes

Make a preliminary visit to the local tax office enquiry counter for their publications:

IR 14/15	*Construction Industry Tax Deduction Scheme*
IR28	*Starting in Business*, and, if needed,
IR40	*Conditions for getting a sub-contractor's tax certificate*
IR53	*PAYE for Employers* (if you employ someone).
IR56/N139	*Employed on self-employed.*

The onus is on the taxpayer to notify the Inland Revenue that he is in business. Failure to do so may result in the imposition of interest and penalties. Either send a letter or use the form provided in the middle of the *Starting in Business* booklet.

Point of contact: telephone directory for address.

VAT Office

Registration for VAT is required if:

1. At the end of any month the value of taxable supplies in the past 12 months has exceeded the annual threshold;

2. There are reasonable grounds for believing that the value of taxable supplies inj the next 30 days will exceed the annual threshold.

Taxable supplies include any zero-rated items. From 1 December 1993 the annual threshold is £45,000. Failure to register is an offence punishable by the imposition of financial penalties.

The VAT office also carries a number of useful publications including;

700	*The VAT Guide*
700/1	*Should I be Registered for VAT*
700/12	*Filling in Your VAT Return*
700/21	*Keeping Records and Accounts*
708/2	*Application of VAT to the Construction Industry*
731	*Cash Accounting*
732	*Annual Accounting*
742	*Land and Property*

Notes on the application of VAT to the Construction Industry are in Chapter 4 together with information about the new 'Cash accounting scheme' and the introduction of annual VAT returns.

Point of contact: telephone directory for address.

DSS Office

Class 2 contributions payable by the self-employed may be paid either in cash or by direct debit through a bank account. Call at the local office to make the necessary arrangements. Class 4 contributions which are also payable by the self-employed are collected along with the income tax by the Inland Revenue and no special action by the businessman is required.

Ask at the DSS office for the following publications:

NI 41	*NI guide for the self-employed*
NI 27A	*People with small earnings from self-employment*
NI 35	*NI for Company Directors*
NI 225	*Direct debit - the easy way to pay* and for employers
NP 15	*Employers' guide to NIC*
NI 227	*Employers' guide to Statutory Sick Pay*

Point of contact: telephone directory for address.

Local authorities

Authorities vary in the provisions made for small businesses but all have been asked to simplify and cut delays in planning applications. In Assisted Areas and Enterprise Zones rent-free periods and reductions in rates may be available on certain industrial and commercial properties. As a preliminary to either purchasing or renting business premises the following booklets will be very helpful. *A Step by Step Guide to Planning Permission for Small Businesses* and *Business Leases and Security of Tenure* are both issued by the Department of Employment and are available at council offices, Citizens' Advice Bureaux and TEC offices.

Some authorities run training schemes in conjunction with local industry and educational establishments.

Point of contact: usually the planning department - ask for the Industrial Development or Economic Development Officer.

Department of Trade and Industry

The current package of assistance from this department is called the Enterprise Initiative and is geared more to the needs of existing businesses than new start-ups. It ranges widely over all aspects; marketing, design, quality, management, finance, etc., and consists essentially of assessing the requirements of the business. The counsellor will keep an eye out for untapped resources, inefficient work systems and unrealized potential and will recommend specialist consultants to come in and advise. The department would pay all costs of the initial survey and one half (or two thirds in development areas) of the costs of any consultancy between 5 and 15 man days.

Point of contact: phone 071-215 5000 and ask for the address and phone number of the nearest DTI office and copies of their explanatory booklets.

Department of the Environment

From 1 April 1992 new regulations are in force under the Environmental Protection Act 1990 relating to all forms of waste other than normal household rubbish. Any concern which produces, stores, treats, processes, transports, recycles or disposes of such waste has a 'duty of care' to ensure it is properly discarded and dealt with. Practical guidance on how to comply with the law (it is a criminal offence punishable by a fine not to) is contained in a booklet *Waste Management: The Duty of Care: A Code of Practice*, obtainable from HMSO Publications Centre, PO Box 276, London SW8 5DT. Phone 071-873 9090. Price £5.00.

Accountant

The services of an accountant are to be strongly recommended from the beginning because the legal and taxation requirements start immediately and must be properly complied with if trouble is to be avoided later. A qualified accountant must be used if a limited company is being formed and for all types of business the accountant should be able to give advice on a whole range of business issues from, for example, book-keeping to grant aid, from tax planning and compliance to finance raising and will clearly help in preparing annual accounts.

It is worth spending some time finding an accountant who has other clients in the same line of business and is able to give sound advice particularly on taxation and business finance and is not so overworked that damaging delays in producing accounts are likely to arise. Ask other traders whether they can recommend their own accountant. Visit more than one firm of accountants, ask about the fees they charge and how much the production of annual accounts and agreement with the Inland Revenue are likely to cost and how long the work will take. A good accountant is worth every penny of his fees but do not hesitate to challenge him if his service is unsatisfactory.

Solicitor

Many businesses operate without the services of a solicitor but there are a number of occasions when legal advice should be sought. In particular no-one should sign the lease of premises without asking a solicitor what they are committing themselves to because it is not unusual for a business to be put into financial difficulty through unnoticed liabilities in its lease. Either an accountant or solicitor will help with drawing up a partnership agreement which all partnerships should have.

A solicitor will also help to explain complex contract terms and prepare draft contracts if the type of business being entered into requires them.

Insurance broker

Policies are available to cover many aspects of business including:

employer's liability - compulsory if the business has employees;
public liability - essential in the construction industry;
motor vehicles;
theft of stock, plant, money, etc;
fire and storm damage;
personal accident and loss of profits.

Brokers are independent advisers who will obtain competitive quotations on your behalf. See more than one broker before making a decision - their advice is normally given free and without obligation.

Point of contact: telephone directory or write for a list of local members to

The British Insurance Brokers Association
Consumer Relations Department
BIBA House
14 Bevis Marks
London EC3A 7NT (Tel: O71-623 9043)

or contact

The Association of British Insurers
51 Gresham Street
London EC2V 7HQ (Tel: 071-600 3333)

who will supply free, a package of very useful advice files specially designed for the small business.

The Health and Safety Executive

The Executive operates the legislation covering everyone engaged in work activities and the following free literature is available:

HSE 16	*The Law on Health and Safety at Work*
IND(G)14(L)	*Compliance with health and safety legislation at work.*
HSE4	*Short guide to Employer's Liability (Compulsory Insurance) Act 1969.*

The Executive has issued a very useful set of 'Construction Health Hazard Information Sheets' covering such topics as handling cement, lead and solvents, safety in the use of ladders, scaffolding, hoists, cranes, flammable liquids, asbestos, roofs and compressed gases, etc. A pack of these may be obtained free from your local HSE office or:

The Health & Safety Executive
St. Hugh's House
Stanley Precinct
Bootle
Merseyside L20 3QY (Tel: 051-951 4381)

FINANCE

Working out how much will be needed

The businessman should estimate in advance what funds will be needed to start up and to run the business for at least the first 12 months. If the forecast is to be reasonably accurate he must make some early decisions about:

1. The premises where the business will be based, what initial repairs, alterations, etc., to them are required and what will be the total cost;

2. What plant, equipment and transport is needed, whether it is to be leased or purchased and again what the cost will be;

3. How much stock of materials, if any, should be carried (the bare minimum only should be acquired so a reliable supplier should be sought out);

4. What will be the weekly bill for overheads, wages and the proprietor's living costs and how these are to be met until the cash for work done starts to flow in;

5. What type of work is going to be undertaken, how much profit margin can realistically be obtained and how often invoices are to be presented.

The above are just some of the items that should be covered. If the proprietor is fortunate enough to have some capital, or some is available within the family, he might get by without doing more than a rough estimate based on his initial research and knowledge of the trade. If however there is the need to seek finance from a bank or other financial institution a much more detailed statement will be required for which the help of an accountant or TEC adviser should be sought.

Sources of funds

Finance, like charity, often begins at home and the would-be businessman should make a realistic assessment of his net worth including the value of his house after deducting any mortgage(s) outstanding on it, his savings, any car or van owned and any sums which his family are prepared to contribute and deduct any private borrowing which will come due for payment in the next 24 months. The whole of these funds may not of course be available (for instance, money which has been loaned to a friend or relative who is known to be unable to repay at the present time). It may not be desirable that it should all be put at risk on a business venture. Establish therefore (a) how much cash it is proposed to invest in the business and (b) whether the family home will be made available for any business borrowings.

Whilst it may be wise not to pledge too much of the family assets, it has to be remembered that the bank will be looking closely at the degree to which the proprietor has committed himself to the venture and will not be impressed by an applicant for a loan who is prepared to risk only a small fraction of his own resources.

Having decided how much of his own funds to contribute, the businessman can now see the level of shortfall and consider how best to fill it. Consideration should be given to partners where the shortfall is large and particularly when there is a need for heavy investment in fixed assets such as premises and capital equipment. It may be worthwhile starting a limited company with others also subscribing capital and to allow the banks to take security against the book debts. The first outside source of money to which most businessmen turn is the bank and a book could be written solely on the dos and don'ts of approaching a bank manager. Here are a few tips:

1. Have a proper business plan to present to him including a cash flow forecast (12 months is usual), also an opening statement of affairs, projected profit and loss accounts and balance sheets for two years and a written statement describing the whole business venture. Use conservative estimates which tend to understate rather than overstate the forecast sales and profits.

2. Know the figures in detail - and don't leave it to an accountant to explain them for you. The bank manager is interested in the businessman not his advisers and will be impressed if he has a sound grasp of the financing of his business.

3. Understand the difference between short and long term borrowing; know how much is needed of each and be ready to explain how the business will be able to repay the bank its money.

4. Ask about the Government Loan Guarantee Scheme if there is a shortage of security for loans. Under this scheme the Government guarantees 70% of the loan up to £100,000 (now up to 85% for loans to businesses in the Inner City Task Force Areas) but if you have been in business for more than two years the limit is £250,000. In return there is an insurance of 0.5% on fixed rate interest arrangements and 1.5% on variable bank interest arrangements. Repayment is over a period from 2 to 7 years.

There are a number of other financial institutions in the 'venture capital' market that can help well established businesses, usually limited companies, wishing to expand and also for some well conceived start-ups. They will provide a flexible package of equity and loan capital in the range of £25,000 to £1,000,000. Usually the deal entails the institution having a minority interest in the voting share capital and a seat on the board of the company. Arrangements for the eventual purchase of the shares held by the finance company by the private shareholders are also normally incorporated in the scheme.

Contact points for information and advice are bank managers, accountants and the TEC. If the outside investor in the business is an individual he will probably wish to invest within the terms of the Enterprise Investment Scheme which enables him to get tax relief at 20% on the amount of his investment. The rules are complex and professional advice is essential.

The Royal Jubilee and Prince's Trusts

These trusts, through the Youth Business Initiative, provide bursaries of not more than £1,000 per individual to selected applicants who are unemployed and aged 25 or under. Grants may be used for tools and equipment, transport, fees, insurance, instruction and training, but not for working capital, rent and rates, new materials or stock. They operate through a local representative whose name and address may be ascertained by contacting The Prince's Youth Business Trust, 5 Cleveland Place, London SW1Y 6JS: Phone 081-968 3713.

THE CONSTRUCTION INDUSTRY TAX DEDUCTION SCHEME

General

The Construction Industry Tax Deduction Scheme is known universally as the '714' Scheme, after the number of the official form around which the whole system revolves. The government is proposing important changes to the scheme but they are still under discussion.

The businessman should visit his local income tax enquiry office and obtain copies of the Revenue booklet IR 14/15 and leaflet IR 40 which explain the conditions under which the Revenue will issue a 714 certificate.

The scheme operates whenever a 'contractor' makes a payment to a 'sub-contractor'.

If the sub-contractor does not hold a valid tax certificate (714I, 714P, 714C or 714S) issued to him by the Inland Revenue then the contractor *must* deduct 25% tax from the whole of any payment made to him (excluding the cost of any materials). If, however, he holds such a certificate the payment may be made in full without deducting tax (in the case of a 714S there is a weekly limit after which tax is deductible).

A business is not obliged by law to seek an exemption certificate and can legally work in the construction industry without one. As a matter of practice, however, many main contractors are reluctant to undertake the additional paperwork required when tax has to be deducted and accounted for to the Revenue and will give work to a sub-contractor with a 714 certificate in preference to one without.

A small business that does work only for the general public and small commercial concerns, is outside the scheme and does not need a 714 certificate to trade. If, however, it engages other contractors to do jobs for it, the business would have to register under the scheme as a contractor and deduct tax from any payments made to a sub-contractor who did not produce a valid 714 certificate.

Obtaining a 714 certificate

There is a special application form which may be obtained from tax offices. There are a number of conditions which have to be met before the Revenue will issue a certificate. In general terms these are:

1. The applicant must be working as a sub-contractor in the UK in the construction industry;

2. The business must be run from proper premises with the usual business facilities, and records must be kept and a business bank account operated;

3. The applicant must have been employed or self-employed in the UK for a continuous period of 3 years in the 6 years before the date of application; short breaks in employment will not be taken into account unless they exceed 6 months in total (but see 5);

4. The applicant must have a satisfactory tax and NIC record: (the Inland Revenue will check the NI contribution position with the DSS);

5. School leavers who can show that in the 6 years up to the date of application they were in full-time education or training for a continuous period of three years, or in full-time education or training for part of the three years and unemployed or in self-employment for the rest, may apply for a special certificate (714S);

6. Those who satisfy all the conditions except 3. above may also obtain a special certificate (714S) if they can arrange for a bank to guarantee the tax payable on the amounts received in full (Leaflet IR 40 sets out the conditions in details).

The 714 certificate

There are four types:

714I - which is issued to individuals;
714P - which is issued to partners;
714C - which is issued to most limited companies;
714S - which is issued as explained above.

The I, P and S certificates include a photograph of the individual to whom it is issued in addition to his name, NIC number, signature and business name (if any). The certificate also has a serial number and an expiry date. Vouchers numbered 715 (in the case of a special certificate, 715S) are issued along with the 714I, P and S.

The 714 system in operation

Holders of certificates 714I and P (including company directors holding 714Ps).

Before paying a sub-contractor in full, without any deduction of tax, the contractor must carry out detailed checks on the 714 certificate (the original, not a copy) and the sub-contractor must produce the 714 for this purpose. On being paid the sub-contractor must complete a voucher 715 and give it to the contractor, showing, among other things, the amount of the payment received. The contractor must send the 715s to the Revenue monthly unless his monthly deduction plus any PAYE and NIC for employees total less than £450, when he may choose to pay quarterly.

Holders of 714C certificates

A sub-contractor which is a limited company may choose to produce to the contractor either the 714C itself or a 'certifying document'. Whichever method is used the documents have to be checked in detail by the contractor, if necessary by telephoning the company itself to confirm that the person presenting the document is their authorized representative.

If the contractor is satisfied he must pay the sub-contractor in full; 715s are not used but both parties record the details of the payment in their business records.

Holders of 714S certificates

If the weekly payment to the sub-contractor (excluding the cost to him of any materials purchased directly from another person) does not exceed the limit shown on the front of the 715S voucher, then the procedure outlined above for a certificate 714I and P is followed and a voucher 715S is given to the contractor on receipt of the payment.
If, however, the payment exceeds the amount shown for the week then the 714I and procedure is followed in respect of the amount shown and the procedure below is carried out for the balance of the payment.

Sub-contractors with no certificate and those with S certificate payments in excess of the weekly limit.

The contractor is obliged to deduct tax from all payments (excluding the cost of directly purchased materials) and to account to the Revenue for all amounts so withheld. To enable the sub-contractor to prove to the inspector of taxes that he has suffered this tax deduction the contractor must give him a certificate on form SC60 showing the amount withheld. These SC60 forms must be carefully filed for production to the Inspector after the end of his accounting year along with his business profit and loss account and balance sheet. Any tax deducted in this way over and above the sub-contractor's proper, agreed liability for the year will be repaid by the Inland Revenue.

At the end of the day the sub-contractor will pay the same amount of income tax whether or not he has a certificate, but those without one will have suffered a severe restriction in their cash flow until the repayment is made.

The main contractor periodically has to send the 715 vouchers to the Inland Revenue Computer Centre in Liverpool and make an annual return to that Centre also. The date for filing the end-of-year return of payments to uncertificated sub-contractors has been extended by one month to 19 May. It all amounts to a heavy burden on the trader with penalties awaiting those who are sloppy, dilatory or dishonest in its operation.

Miscellaneous points

1. A payment includes anything paid out by the contractor such as a 'sub' or a loan, whether by cash, cheque or credit and whether direct to the sub-contractor or to his nominee.

2. The cost of subsistence and travelling expenses reimbursed by the contractor is included in the amount on which the tax deduction is calculated.

3. The scheme is policed by the Revenue in much the same way as they inspect PAYE documents, and records for the scheme have to be made available on request.

4. A contractor is liable to pay to the collector of taxes all amounts which he *should* have deducted from sub-contractors, whether he made the deductions or not. He may, however, be excused from having to pay if he can show he took reasonable care and made the error in good faith.
   If the deductions have not been properly made the sub-contractor himself will be asked to account for his own correct liability.

5. In a situation where tax *has* been deducted from payments made by him but the contractor has failed to account for it to the collector, the sub-contractor would be unlikely to recover any over-deductions from the Revenue.

6. Disputes between contractors and sub-contractors about the amount of any deductions should be referred to the inspector for a ruling.

Note

The above is merely a summary of the very detailed instructions contained in IR 14/15 paragraphs 65-129 which must be carefully studied by anyone involved in operating the scheme either as a contractor or sub-contractor.

VAT

The general rule about liability to register for VAT is given in the VAT office notes above. It is possible to give here only a brief outline of how the tax works. The rules which apply to the construction industry are extremely complex and all traders must study *The VAT Guide* and other publications.

The amount of tax to be paid is the difference between the VAT charged out to customers *(output tax)* and that suffered on payments made to suppliers for goods and services *(input tax)*. Unlike income tax there is no distinction in VAT for capital items so that the tax charged on the purchase of, for example, machinery, trucks and office furniture, will normally be reclaimable as Input Tax. One important exception to this is that the input tax on a car cannot be reclaimed even though it is used wholly for business purposes.

VAT is payable in respect of 3-monthly periods known as 'tax periods' and you can apply to have the group of tax periods which fits in best with your financial year. The tax must be paid within one month of the end of each tax period. Traders who receive regular repayments of VAT can apply to have them monthly rather than quarterly.

Not all types of goods and services are taxed at 17.5% (i.e., at the standard rate), some are exempt and others are zero-rated.

Zero-rated

This means that no VAT is chargeable on the goods or services but a registered trader can reclaim any **input** tax suffered on his purchases. For instance a builder pays VAT on the materials he buys, but if he is constructing a new dwelling house, this is zero-rated and he may reclaim this VAT or set it off against any VAT due on other standard rated work.

Exempt

Supplies which are exempt are less favourably treated than those which are zero-rated. Again no VAT is chargeable on the goods or services but the trader cannot reclaim any **input** tax suffered on his purchases.

Standard Rated

All work which is not specifically stated to be zero-rated or exempt is standard rated, i.e., VAT is chargeable at the current rate of 17½% and the trader may deduct any **input** tax suffered when he is making his return to the Customs and Excise.

If for any reason a trader makes a supply and fails to charge VAT when he should have done so (e.g. mistakenly assuming the supply to be zero-rated) he will have to account for the VAT himself out of the proceeds. If there is any doubt about the VAT position it is safer to assume the supply is standard rated, charge the appropriate amount of VAT on the invoice and argue about it later.

Time of supply

The *time* at which a supply of goods or services is treated as taking place is important and is called the 'tax point'. VAT must be accounted for to the Customs and Excise at the end of the accounting period in which this 'tax point' occurs. For the supply of *goods* which are 'built on site' the 'basic tax point' is the date the goods are made available for the customer's use, whilst for *services* it is normally the date when all the work except invoicing is completed.

However, if you issue a tax invoice or receive a payment *before* this 'basic tax point' then that date becomes the tax point.

In the case of contracts providing for stage and retention payments the tax point is either the date the tax invoice is issued or when payment is received, whichever is the earlier.

All the above requirements apply to sub-contractors and main contractors and it should be noted that, when a contractor deducts income tax from a payment to a sub-contractor (because he has no valid 714), VAT is payable on the full gross amount *before* taking off the Income Tax (see Chapter 3).

See below for examples of VAT payments and repayments.

Examples of how VAT works are as follows:

(a) Sales invoices total	£1000
plus VAT @ 17½%	175
The customer pays	£1175
Purchase invoices total	£600
plus VAT @ 17½%	105
Total paid	£705

In this example the *output* tax is £175 and the *input* tax is £105 which means that the trader owes £70 to the Customs and Excise.

If in the tax period machinery costing £2000 plus VAT £105 had also been purchased the figures would be:

(b) Sales invoices total	£1000
plus VAT @ 17½%	175
The customer pays	£1175

Purchase invoices total	£2600
plus VAT @ 17½%	455
Total paid	£3055

In this example the *output* tax is also £175 but the *input* tax is £455 so that a refund of £280 is due to the trader from the Customs and Excise.

Chapter 2
Running the business

Many businesses are run without adequate information being available to check trends in their vital areas, e.g. marketing, money and managerial efficiency. It is vital to look critically at all aspects of the business to maximize profits and eliminate inefficiency.

Proprietors often have the feeling that the business should be 'doing better' than it is, without being able to identify what is going wrong. Sometimes there is the worrying phenomenon of a steadily increasing work programme coupled with a persistently reducing bank balance or rising overdraft.

Some useful ways of checking the position and identifying problem areas are given below.

Marketing

Whilst management and finance are concerned with the internal running of a business the market is where it makes contact with the outside world in the shape of its competitors and customers. Throughout his business life the entrepreneur should study carefully the methods and approach of the former and the needs and wishes of the latter. A shortcoming frequently found in ailing concerns is that the proprietor thinks he knows better than his customers what they want.

The term 'market research' sounds both difficult and expensive but a very simple form of it can be done quite effectively by the businessman and his sales staff. First, identify the type of person or business to whom the products or service are likely to appeal, finding out and recording what it is the customer wants in terms of price, quality, design, payment terms, follow-up service, guarantees, services.

The initial approach might be by leaflet or letter followed by a personal call. As an on-going part of management all staff with customer contact should be encouraged to enquire about and record customer preferences, complaints, etc., and feed it back to management.

Other sources of information are friends in the trade, business journals, trade exhibitions, suppliers, representatives, etc., from whom information about trends, new techniques and products can be obtained and studied.

Valuable information can also be gained from studying competitors and the following questions should be asked:

- what do they sell and at what prices?

- what inducements to buy do they offer their customers (e.g. credit facilities, guarantees, free offers, discounts, etc.)?

- how do they reach their customers (local/national advertising; mail shots; salesmen; local radio and TV)?

- what are the strongest aspects of their appeal to customers and have they any weaknesses?

The businessman should apply all the information gathered from customers and competitors to his own range of products with a view to making sure he is offering the right product at the right price in the most attractive way and in the most receptive market.

In a small business where the proprietor is also his own salesman he must give careful thought to how he can best present his product and himself. If he is working solely within the construction industry his main problems are likely to centre on getting a 714 and on exploiting as fully as possible trade contacts to get sub-contract work.

However, for those who serve the general public, presentation can be a vital element in getting work. The customer is looking for efficiency, reliability and honesty in a trader and quality, price and style in the product. To bring out these facets in discussion with a potential customer is a skilled task and for newcomers to business and those who have not had as much success to date as they hoped for, a short course on marketing techniques could pay handsome dividends.

The Enterprise unit of the Manpower Services Commission will give the names and addresses of such courses locally - contact it through the job centre or make an appointment to see a TEC adviser with marketing experience and talk over the problems with him.

Some indications that a market review is needed are:

- declining sales

- profit margins being squeezed

- fewer customers and over-reliance on a few large ones

- profitability not spread evenly over the product or service range - some items 'not earning their keep' - perhaps through being out of date, too expensive, badly designed, poorly constructed, etc.

Unfortunately some firms which close down do not seek financial advice until it is too late to halt the downward trend when earlier attention to the problems may have saved some of them. There are many reasons for this and one of them is that those running the business are unable to recognize the tell-tale signs and very few accountants take the trouble to explain to their clients what to look for. There are some tests and checks that can be done quite easily.

Cashflow

Cashflow is the lifeblood of the business and more businesses fail through lack of cash than for any other reason. Cash is generated through the conversion of work into debtors and then into payment and throught the deferral of the payment of supplies for as long a period as can be negotiated.

The objective must be to keep stock, work in progress and debts to a minimum and creditors to a maximum. Trends in important ratios as well as absolute values can help in assessing the business performance.

Debtor days

This is calculated by dividing your trade debtors by annual sales and multiplying by 365. It shows the number of days' credit being afforded to your customers and should be compared both with your normal trade terms and the previous month's figures. Normal procedures should involve the preparation of a monthly aged list of debtors showing the name of the customer, the value and which month it relates to.

The oldest and largest debtors can be seen at a glance for immediate consideration of what further recovery action is needed. The list may also show over-reliance on one or two large customers or the need to stop supplying a particularly bad payer until his arrears have been reduced to an acceptable level. Consideration should be given to making up bills to a date before the end of the month and making sure the accounts are sent out immediately followed by a statement 4 weeks later. Consider giving discounts for prompt payment.

If all else fails, and legal action for recovery is being contemplated, call at the County Court and ask for their leaflets, numbers 1 to 4.

Stockturn

The level of stock should be kept to a minimum and the number of days' stock can be calculated by dividing the stock by the annual purchases and multiplying by 365. A worsening trend on a month by month basis shows the need for action. It is important to make regularly a full inventory of all stock and dispose of old or surplus items for cash. A stock control procedure to avoid stock losses and to keep stock to a minimum should be implemented.

Profitability

Whilst cash is vital in the short-term, profitability is vital in the medium-term. The two key percentage figures are the gross profit percentage and the net profit percentage. Gross profit is calculated by deducting the cost of materials and direct labour from the sales figure whilst net profit is arrived at after deducting all overheads.

Possible reasons for changes in the gross profit percentage are:

- the pricing of jobs is not taking full account of increases in materials and wages and a review of pricing policy is needed

- too generous discount terms are being offered

- poor management, overmanning, waste and pilferage of materials

- too much down-time on plant which is in need of replacement.

If net profit is deteriorating after the deduction of an appropriate reward for your own efforts, including an amount for your own personal tax liability, you should review each item of overhead expenditure in detail asking amongst others the following questions:

- can savings be made in non-productive staff?

- is sub-contracting possible and would it be cheaper?

- have all possible energy-saving methods been fully explored?

- do the company's vehicles spend too much time in the yard; can they be shared and their number reduced?
- is the expenditure on advertising producing sales (review in association with 'marketing' above)?

Over trading

Many inexperienced businessmen imagine that profitability equals money in the bank. In some cases, particularly where the receipts are wholly in cash, this may be the case, but additional business may mean higher stock inventories, extra wages, overheads, increased capital expenditure on premises and plant.

If the debtors show a marked increase as the turnover rises the proprietor may find to his surprise that each expansion of trade reduces rather than increases his cash resources. The business, which had enough funds for start-up, finds it does not have sufficient cash to run at the higher level of operation and the bank manager may be getting anxious about the increasing overdraft.

It is essential for those who run a business which operates on credit terms to be aware that profitability does not necessarily mean increased cash availability. Regular monthly management information on marketing and finance as described in this chapter will enable 'over trading' to be recognized and remedial action to be taken early.

If the situation is appreciated only when the bank and other creditors are pressing for money, radical solutions may be necessary such as bringing in new finance, sale and leaseback of premises, a fundamental change in the terms of trade, or even selling out to a buyer with more resources. Professional help from the firm's accountant will be needed, whilst the TEC has counsellors experienced in advising on both the marketing and financial aspects of such situations.

Break-even point

The costs of a business may be divided into two types: variable and fixed.

Variable costs are those which increase or decrease as the volume of work goes up or down and include such items as materials used, direct labour, power, machine tools, etc.

Fixed costs are not related to turnover and are sometimes called 'fixed overheads'. They include rent, rates, insurance, heat and light, office salaries, plant depreciation, etc; these costs are still incurred even though few or no sales are being made.

Many small businessmen run their enterprises from home using family labour as back-up; they sell mainly their own labour and buy materials and hire plant only as required.

By these means they reduce their fixed costs to a minimum and start making profits almost immediately. However, larger firms which have business premises, perhaps a small workshop, an office, vehicles, etc., need to know how much they have to sell to cover their costs and become profitable.

In the case of a new business it is necessary to estimate the figures but where annual accounts are available a break-even chart based on them can be readily prepared.

Suppose the real or estimated figures (expressed in £000s) are:

	%	£000
Sales	100	400
Variable costs	66	265
Gross profit	34	135
Fixed costs	13	50
Nett profit	21	85

$$\text{Break-even point} = \text{Fixed Costs} \div 1 - \frac{\text{Variable Costs}}{\text{Sales}}$$

$$= 50 \div 1 - \frac{265}{400}$$

$$= 50 \div (1 - 0.6625)$$

$$= 50 \div 0.3375$$

$$= £148 \text{ (thousand)}$$

In practice things are never quite as clear cut as the figures above show, but nevertheless this is a very useful tool for assessing not only the break-even point but also the approximate amount of loss or profit arising at differing levels of turnover and also for considering pricing policy.

Chapter 3
Taxation

A new basis of charging tax on business profits took effect from 6 April 1994 for new businesses and from 6 April 1996 for businesses already established at 6 April 1994.

For new businesses there may still be an advantage in having an accounting date early in the tax year, that is shortly after 6 April. The advantage, however, is less than it would have been under the old rules and professional advice will be needed to choose the best accounting date for tax purposes.

For existing businesses at 6 April 1994 the way income tax is charged on business profits is to change radically. For the 1997/98 tax year onwards profits assessed for tax in a given tax year will be the profits of the accounting period ending in that year and not as now those of the preceding year.

Transitional rules will apply for the tax year 1996/97 whereby the profits of two accounting periods will be used to give an average. For example, a self-employed business with 30 April year end, the average period began on 1 May 1994.

Because of the averaging of profits charged to tax, a tax planning opportunity will arise subject to anti-avoidance provisions introduced by the Inland Revenue to penalize the artificial movement of profit into the averaging period.

In particular the following areas may offer tax planning opportunities:

1. the purchase as opposed to the leasing of capital equipment to be used in the business

2. the introduction of a partner into the business

3. where a partnership is considering for sound business reasons, re-financing by way of personal loans rather than a partnership loan.

Professional advice will be needed to take the maximum advantage of these opportunities.

The timing of a cessation of business will still be important particularly until the new system is fully in place. Professional advice should be sought on the timing and tax cost of ceasing business.

Together with changes in the way taxable profits will be measured, the Inland Revenue are introducing a new system of charging and collecting personal tax. Again from the tax year 1996/97 the burden of assessing tax will shift from the Inland Revenue to the individual tax payer. The main features of this new system are as follows:

1. the onus is on the taxpayer to provide and complete information

2. the taxpayer will have a choice: he can calculate and pay his tax liability at the same time as making his return and this will need to be done by 31 January following the end of the tax year. Alternatively he can send in his tax return much earlier and the Inland Revenue will calculate the tax to be paid on 31 January

3. the important aspect to the new system is that if the return is late, or the tax is paid late, there will be automatic penalties, imposed on the taxpayer.

Spouses in business

If spouses work in a business, perhaps answering the phone, making appointments, writing business letters, making up bills and keeping the books, they should be properly remunerated for it. Being a payment to a family member the inspector of taxes will be understandably cautious in allowing it in full as a business expense. The payment should be:

1. actually paid to them, preferably weekly or monthly and in addition to any housekeeping monies

2. recorded in the business book

3. reasonable in amount in line with their duties and the time spent on them.

If the wages paid to them exceed £56.99 p.w. Class 1 employer and employee NIC becomes due and if they exceed £3,445 p.a. (assuming they have no other income) PAYE tax will also be payable.

It should also be noted that once small businesses are well established and the spouse's earnings are approaching the above limits, consideration may be given to bringing them in as a partner. This has a number of effects:

1. there is no longer a need to relate the spouse's income (which is now a share of the profits) to the work they do (if any)

2. they will pay Class 2 and Class 4 NIC instead of the more costly Class 1 contributions and PAYE will no longer apply to their earnings

3. but remember - as partners any assets they own are vulnerable to proceedings by partnership creditors.

Premises

Many small businessmen cannot afford to rent or buy commercial premises and run their enterprises from home using part of it as an office where the books and vouchers, clients' records, trade manuals, etc. are kept and estimates and plans are drawn up. In these circumstances a portion of the outgoings on the property may be claimed as a business expense.

Car expenses are usually split on a fractional mileage basis between business journeys, which are allowable, and private ones, which are not, and a record of each should be kept. If the business does work only on one or two sites or for only one main contractor the inspector may argue that the true base of operations is the work site not the residence and seek to disallow the cost of travel between home and work. It is tax-wise therefore, and sound business practice, to have as many customers as possible and not work for just one client.

Appeals against assessments. To businessmen, income tax assessments are an anathema. They should resist the temptation to tear them up or put them behind the clock and forget about them. All assessments should be checked for accuracy immediately and, if excessive, the instructions on the notice about making an appeal should be followed and the appeal sent to the Tax District that issued the assessment. The appeal should also show how much of the tax charged should be postponed because the assessment is too high.

If this is not done within 30 days of the issue of the notice, the assessment becomes final and the inspector (and the general commissioners if they are asked to adjudicate) may well not accept a late appeal. In this event the taxpayer has no alternative but to pay the tax as charged on the assessment which may be estimated or contain additions to the profits which have not been agreed by him and his accountant. If the appeal is to be made by the accountant check with him before the 30 days are up to ensure that it has been submitted.

Keep copies of all correspondence with the inspector and collector. Letters can be mislaid or fail to be delivered and it is essential to have both proof of what was sent as well as a permanent record of all correspondence.

Some useful statistics on Income Tax 1994/95

The current personal allowance for a single person is £3,445. The personal allowances for people aged 65-74 and over 75 years are £4,200 and £4,370 respectively. The married couple's allowance is £1,720 and £2,665 for a couple between the ages 65 to 74 and £2,705 for a couple over 75 years.

Rates of tax

The value of the married couple's allowance has been reduced for 1994/95 as it is now only worth a maximum 20% instead of saving tax at a person's top rate. The rates of tax for 1994/95 are as follows:

Lower rate: 20% on taxable income up to £3,000
Basic rate: 25% on taxable income between £3,000 and £23,700
Higher rate: 40% on taxable income over £23,700

Example: Married man: wife has no income. On earnings of £20,000 he would pay:

		£
Income		20,000
Less personal allowance		3,445
Taxable income		16,555
	£3,000 at 20% =	600
	£13,555 at 25% =	3,389
		3,989
Less married couple's allowance	£1,720 at 20%	344
Tax payable		3,645

On earnings of £30,000 the tax he would pay:

	£3,000 at 20% =	600
	£20,700 at 25% =	5,175
	£2,855 at 40% =	1,142
Less married couple's allowance	£1,720 at 20%	344
	Tax payable =	6,573

Taxation of husband and wife

A married woman is treated in much the same way as a single person with her own personal allowance and basic rate band. Husband and wife each make a separate return of their own income and the Inland Revenue deals with each one in complete privacy - letters about the husband's affairs will be addressed only to him and about the wife's only to her (unless the parties indicate differently). The allowances and relief are dealt with as follows.

Personal allowance

Husband and wife each has one of these - for 1994/95 it is £3,445 each.

Married couples allowance

The amount is £1,720. This is initially due to the husband but if his income is too small to use it all he may transfer the surplus part to his wife but once he has passed some allowances in this way he cannot change his mind and ask for them back. Married couples may choose, however, how they wish the allowance to be allocated between them, subject to the right of the wife to claim half the allowance if she wishes. Here are two examples:

Husband and wife both in employment; both earn £7,000pa.

1993/94: husband decides to keep all the married couples allowance. The relief would be:

	Husband	Wife
	£	£
Personal allowance	3,445	3,445
Married couples allowance	1,720	-
	£ 5,165	£3,445

1994/95: wife decides she wants her half share of the allowance The relief would be:

	Husband.	Wife.
	£	£
Personal Allowance	3,445	3,445
Married couple's allowance	860	860
	£4,305	4,305

Basic rate band

Husband and wife each have £3,000 chargeable at 20% and up to £20,000 at 25%.

Mortgage interest relief

The ceiling for relief is unchanged for 1994/95 at £30,000. The value of relief, however, has been reduced in a similar way to married couple's allowance in that there is a maximum rate of relief of 20%. If the loan is in the name of one spouse only, that one gets all the relief due. If it is in the joint names of husband and wife it is allowed equally between them. However if both agree they can *jointly* ask for relief to be divided in any proportion that they wish.

Business losses

These are allowed only against the income of the person who incurs the loss. For example, a loss in the husband's business cannot be set against the wife's income from employment.

Joint income

In the case of joint ownership by husband and wife of assets which yield income bank and building society accounts, shares, rented property, etc. - the Revenue will treat the income as arising equally to both and each will pay tax on one half of the income. If however the asset is owned in unequal shares or by one spouse only and the taxpayer can prove this, then the shares to be taxed can be adjusted accordingly if a joint declaration is made to the tax office setting out the facts.

General

Special rules apply in the year of marriage or separation and divorce and on the death of the husband or wife. Contact your local tax office for information.

Capital Gains Tax (individuals 1994/95)

Where an asset is disposed of, the first £5,800 of the gain is exempt from tax. In the case of husbands and wives, each has a £5,800 exemption so if the ownership of the assets is divided between them, it is possible to claim exemption on gains up to £11,600 jointly in the tax year. Any remaining gain is chargeable as though it were the top slice of the individual's income, therefore, according to his or her circumstances it might be charged at 20%, 25% or 40%. Here is an example:

Husband: He is in business and is liable at 40% on his profits. He has £7,000 of chargeable capital gains less the £5,800 exemption sum which leaves £1,200. This is taxed at 40% to produce a sum due to the Inland Revenue of £480.

Wife: She has no income but also has £7,000 of chargeable gains. Her exemption of £5,800 leaves a taxable sum of £1,200 and as she has no taxable income and the gain is less than the lower band it is charged at 20% making £240 to be paid in tax.

Retirement relief may be due on the disposal of certain business assets after the age of 55 (or before that date where retirement is due to proven ill health). The maximum relief against capital gains is £250,000 plus one half of the gains between £250,000 and £1,000,000. A businessman contemplating retirement, or sale of business when aged 55 or over should consult his accountant *before* taking any steps and *before* changing his working pattern (e.g. going part-time).

Business entertainment

No relief is due for expenditure on business entertainment or on gifts to customers, whether they are from this country or overseas. However, the cost of small trade gifts not exceeding £10 per person in value is still admissible provided that the gift advertises the business and does not consist of food or tabacco.

Construction industry

An uncertificated sub-contractor in the construction industry will suffer tax at the basic rate of 25%. Any overpayment on account of the 20% rate may be reclaimed at the end of the tax year.

Dates tax due

Income Tax 1994/95

Earned income (such as trading profits) 50% on 1 January 1995 and a further 50% on 1 July 1995. Unearned income (such as rents and interest) due 1 January 1995.

Capital Gains Tax

Tax on gains in year ended 5 April 1994 is due on 1 December 1994 (or within 30 days of the issue of the notice of assessment). Tax on gains arising in the current tax year are due on 1 December 1995.

Self-employed NIC rates (from 6 April 1994)

Class 2 Rate - £5.65 per week. If earnings are below £3,200 p.a. averaged over the year ask the DSS about 'small income exception'; details are in leaflet NI 27A.

Class 4 Rate - Business profits up to £6,490 p.a.: NIL. Profits between £6,490 and £22,360 p.a.: 7.3% of the profit. There is no charge on profits over £22,360 p.a. so the maximum amount of Class 4 contributions is £1,158.51. One half of the Class 4 contribution is deducted from the profits for income tax purposes. Class 4 contributions are collected by the Inland Revenue along with the income tax due.

Corporation Tax (years ended 31/3/94 and 31/3/95)

For the year ended 31/3/95 Corporation Tax is charged at 25% for profits up to £300,000. The ceiling for the previous year ended 31/3/94 was £250,000. Where the accounting date of a company is not 31/3/94, profits have to be apportioned on the time basis to the respective tax years. Profits exceeding £300,000 will be effectively charged at 35% up to £1,500,000 when the rate reduces to 33%. Companies can carry back trading losses for up to 3 years.

Capital allowances (depreciation) rates

Plant and machinery	25%	
Business motor cars - cost up to £12,000	25%	
- cost over £12,000	£3,000	(maximum)
Industrial buildings	4%	
Commercial and industrial buildings in Enterprise Zones	100%	

VAT

The standard rate is 17.5%. Lower limit for registration (from 11 March 1992) turnover per annum - £45,000.

Bad debts and VAT

Relief is available for debts over 6 months.

Chapter 4
Estimating

Pity the poor estimator! If a job goes well there will be a queue of agents, foremen, tradesmen, surveyors and buyers to take the credit. If a job loses money, however, the one person who will be left isolated to take the blame is the unsung hero in the construction industry - the estimator!

His art is highly dependent upon making a series of intelligent guesses to fill in the gaps of information not covered by the specification drawings and/or the bills of quantities. The quality of these guesses (or subjective judgements as the jargon would have it!) will often make the difference between a job making a profit or a loss.

It is possible, of course, to have too much information! There are many examples of a contractor carrying out the first phase of a contract but losing the second phase in open tender despite the fact of having the site set-up already there. The reason for the loss is usually due to having too much local knowledge gained from working on Phase 1 and including the cost of the known local risks in the second tender. Ignorance frequently prevails!

Apart from determining the contract sum at the tender stage, a properly prepared estimate can also be used for the following purposes:

1. Calculation of bonus targets;

2. Preparation of material schedules;

3. Analysis of anticipated and actual costs;

4 Production of a programme;

5. Preparation of monthly valuation and the final account.

This book, however, is only concerned with the preparation of estimates at the tender stage. It should be recognized that estimating is a very imprecise art. This often surprises people outside the construction industry who imagine it to be '.....merely a matter of counting bricks and pricing them' as someone once said to me. If only it was!

Even taking this comment at face value shows how inaccurate this view is. There are many different kinds of bricks all with a different value. The cost of sand and cement for the mortar varies from supplier to supplier as does the hire cost of a mixer. The labour costs are the most likely to show the greatest variation.

No two men work at the same pace and produce the same volume of work. So even an uncomplicated task such as building a wall can produce significantly different estimates. When it comes to more complicated work, the likelihood of different estimates producing wide variations in their tender increases proportionately.

The main divisions in a properly prepared estimate are:

1. Own work which can be sub-divided into
   (a) Labour
   (b) Materials
   (c) Plant;

2. Work to be sub-let to sub-contractors;

3. Site overheads;

4. Office overheads;

5. Profit.

When an enquiry is received a contractor should decide very quickly which parts of the work he would sub-let if he was awarded the contract. He should then send the relevant extracts from the enquiry documents (specification/bills of quantities/drawings) to two or three sub-contractors whilst he is preparing the remainder of the estimate.

Labour

An experienced contractor should know the net charge-out rate of the men he directly employs. The basic labour rate on which the information in this book is based is set out at the beginning of Chapter 5. Every contractor will probably have his own 'customized' version but the main items such as NI contribution, overtime payments, bonus etc must be included.

The particular circumstances of the job must also be considered. The work may have to be carried out outside normal trading hours or done in unpleasant working conditions - both these situations would produce higher labour costs.

Materials

Although it is usually possible to get a better price from another materials supplier, a contractor should weigh up whether it is worth spending a lot of his time investigating other sources of supply if the savings are marginal.

A supplier who delivers on time (including the occasional small item when necessary) and replaces defective materials without query is probably worth supporting even if his prices are slightly higher than some of his competitors. It is still worth checking other prices now and then, however, to make sure that they are not too far out of line; but reliability and service have a real value and should be recognized.

The most difficult aspect of pricing materials is making a realistic assessment for waste and theft. Materials which are bought in bulk and only part of the order is allocated to any particular job or operation usually attract the highest waste percentage which can be as high as 15 to 20%. Specialized 'one-off' items such as cylinders, heaters and the like are usually better looked after and the waste factor could be as low as 1%.

Theft is another problem and a prudent contractor should make some allowance in his estimates to cover for this frustrating part of the industry.

Plant

Most small contractors hire in plant as necessary so the estimator need only assess the time the plant will be required because the hire rate will be easily established. One point to watch out for is the question of minimum hire periods.

Even if a piece of equipment is only required for one hour, the full cost of a day's hire must be included in the rates if that is the minimum hire period that can be obtained.

Site overheads

The range of overheads to be provided on site will vary widely depending on the size and nature of the job. There are two main types of overheads - fixed and time-related. Examples of these will clearly reveal the difference between them.

Fixed overheads are those of a non-recurring nature such as the establishment of the site huts. The cost of this operation - say £500 - must be allowed for whether the job lasts 4 weeks or 104 weeks and is a fixed sum.

Time-related overheads however are totally linked to the time the facility or service is required. In the case of site hutting the estimator should allow a weekly cost for the anticipated contract period plus whatever time he thinks the huts will be needed for carrying out maintenance work.

The following should act as a check list for both fixed and time-related overheads:

Site supervision
Site accommodation
Lighting
Water
Safety, health and welfare
Removal of rubbish
Cleaning
Drying out
Protection of work
Security
Small tools
Insurances
Travelling time and fares
Scaffolding
Temporary services
Temporary fencing and screens

Office overheads

All contractors should give careful thought to the cost of their office overheads and in particular should be aware of the relationship, expressed as a percentage, between these costs and turnover.

The following is an example of a small contractor's overheads. This list is not meant be exhaustive.

				£
Rent			(52 x £50)	2,600
Printing, stationery, etc				150
Telephone			(12 x £50)	600
Van	HP		(12 x £200)	2,400
	Insurance			400
	Tax			100
	Petrol		(52 x 30)	1,560
	Repairs	say		500
Carried forward			£	8,310

Brought forward	£	8,310
Part-time clerk/typist (52 x 30 hours x £3.50)		5,460
Photocopier - lease	(52 x £25)	1,300
Word-processor - lease	(52 x £20)	1,040
Advertising	(52 x £20)	1,040
Membership of professional body		150
Sundries		1,000
	£	18,300

Let us assume that the firm's turnover (that is, the total amount billed for the previous year) is £100,000. A simple calculation of £18,300 x 100 ÷ £100,000 shows that the relationship of overheads to turnover is 18.3%. I have set out a table below showing how this percentage varies when the level of the overheads or the turnover changes.

Overheads £	Turnover £	%
18,300	100,000	18.30
18,300	80,000	22.90
18,300	120,000	15.25
24,000	100,000	24.00
30,000	80,000	37.50
12,000	120,000	10.00

It is important to know the current percentage so it can be added to each quotation (although not necessarily shown). Ask your accountant for the cost of the overheads for last year and assess the percentage to be added to the costs. Unless something dramatic happens mid-year this figure needs only be examined annually. An example which would affect the overheads total could be the employment of a non-working supervisor or the purchase of an extra vehicle. A cancellation of a project or an unexpected increase in work would also affect the turnover figure.

It can be seen, therefore, that although working on a notional percentage is acceptable you should keep your eye open for any circumstances which may make a significant alteration to it.

Profit

Profit is the difference between income and cost. Some newly established contractors find it puzzling that although they are certain that their work is profitable, it is not reflected in the bank balance! The simple explanation for this apparent paradox is that as soon as it is earned the profit is re-used to finance the next stage of work or the next job. It only appears as a tangible asset when either trading stops or after a series of profitable jobs are completed and paid for.

Expansion of the business (and it is very difficult for a successful contractor not to expand) will further delay the surfacing of the profit and even a well planned expansion programme will usually result in an increased overdraft. The profit is still in the business of course, although not in the form of cash in the bank but in debtors accounts and work in progress.

Chapter 5
Rates for measured work

Generally

The rates contained in this chapter are average costs and generally apply to contracts ranging from minor works to those up to £50,000 in value. The rates are based upon national average prices and the following adjustments should be made for working in different regions.

Scotland	0%
Wales	-7%
Northern Ireland	-20%
England	
South West	-10%
South East	-2%
Home Counties	+2%
Inner London	+10%
Outer London	+3%
East Anglia	-4%
Midlands	-8%
North West	-5%
North East	0%

The material prices are based at April 1994 and the labour rates based upon those agreed by The Joint Industry Board for Plumbing Mechanical Engineering Services which took effect from 4 April 1994.

The rates are exclusive of VAT which may be chargeable depending on the nature of the work. The rates are divided into their component parts and this enables any of them to be adjusted individually.

Item descriptions

These contain brief descriptions of items. Item descriptions for fittings are listed so that the more commonly used fittings appear first. Each item includes all the work normally associated with that particular item even if it is not expressly stated, e.g. tank and tap connections include the screwed connections to the tanks and taps respectively. An allowance has been made in the Waste and Sundries column for such items as flux, solder, screws and plugs.

Unit

This column gives the unit in which the item is normally measured. Pipework and gutterwork are usually measured in linear metres designated 'm' and other items are enumerated and marked 'nr'.

Plumber hours

The time taken for the fixing of each of the items is expressed as a decimal fraction of an hour; therefore 0.50 plumbers' hours equals thirty minutes. The times are average and include unloading, distributing and fixing in position, and are based on two advanced plumbers assisted by a third-year apprentice and reduced to a composite rate as illustrated later.

Net Plumber

This column gives the total labour cost of supplying and fixing the item concerned. It is calculated by multiplying the total plumber hours by the hourly rate. The all-in rate is calculated in accordance with the rates of wages agreed and published by the Joint Industry Board for Plumbing, Mechanical Engineering Services in England and Wales as follows:

	Trained plumber £	Advanced plumber £	Technical plumber £
Basic rate (from 4 April 1994)	4.96	5.49	6.20
Basic rate x total hours worked (1910 hrs)	9,473.60	10,485.90	11,842.00
Non-productive overtime (50 hrs)	248.00	274.50	310.00
Public holidays (60 hrs)	297.60	329.40	372.00
Sub-total (1)	10,019.20	11,089.80	12,524.00

RATES FOR MEASURED WORK

	£	£	£
Brought forward	10,019.20	11,089.80	12,524.00
National Insurance (based on sub-total 1)	764.40	861.50	956.00
Annual holiday with pay and sickness benefit 52 weeks at:			
£17.23 (Trained Plumber)	859.96		
£18.76 (Advanced Plumber)		975.52	
£21.01 (Technical Plumber)			1,092.52
Industry pension scheme 6.5% of sub-total (1)	651.25	720.84	814.06
Tool money 45.2 weeks at £2.24	101.25	101.25	101.25
CITB Levy (0.25%)	25.05	27.72	31.31
Sub-total	12,421.11	13,776.63	15,519.14
Severance pay 1.5%	186.32	206.65	232.79
Employer's liability and third party insurance 2%	248.42	275.53	310.38
TOTAL ANNUAL COST	12,855.85	14,258.81	16,062.31
COST PER HOUR divided by 1822 (1911 less 89 = 1822 productive hours)	6.89	7.63	8.59

RATES FOR MEASURED WORK

	Apprentice 1st year £	Apprentice 2nd year £	Apprentice 3rd year £
Basic rate	1.55	2.38	2.94
Basic rate x total hours worked (1574 hrs)	2,439.70	3,746.12	4,627.56
Non-productive overtime (40 hrs)	62.00	95.20	117.60
Public holidays (60 hrs)	93.00	142.80	176.40
Sub-total (1)	2,594.70	3,984.12	4,921.56
National Insurance	-	137.28	225.68
Annual holiday with pay and sickness benefit 40 weeks at £4.74 (1st year) and £7.75 (2nd and 3rd years)	189.60	310.10	310.10
Industry pension scheme 6.5% of sub-total (1)	168.66	258.97	319.90
Tool money 43.2 wks at £2.24	96.77	96.77	96.77
Sub-total	3,049.73	4,787.14	5,873.91
Severance pay 1.5%	45.75	71.81	88.11
Employer's liability and third party insurance 2%	61.00	95.74	117.48
TOTAL ANNUAL COST	£3,156.48	£4,954.69	£6,079.50
Divided by total number of productive hours: 1574 hrs less 40 hrs inclement weather = 1534 hrs	2.06/hr	3.23/hr	3.96/hr

The hourly rate is therefore:

Two advanced plumbers at £7.63	15.26
One 3rd year apprentice	3.96
divided by 2 (number of principal operatives)	19.22
	£9.61 say £9.60

Net material

The material costs are set out in the front of each of the work sections. Generally speaking, the costs reflect the manufacturers' list prices and exclude trade discounts which are usually negotiated by each individual contractor depending upon his creditworthiness, turnover and length of association with the supplier. The prices are based upon the purchase of manufacturers' standard packs. Where items are described as 'Extra over for', the full value of the item is given and no deduction is made for the length of pipe taken up by the fitting. Partly to compensate for this minor overmeasure, no allowance has been made for joining materials. For pipework and gutters, an allowance has been made for brackets, clips, holderbats, etc., based on the assumption that they would be placed every two metres.

In the case of copper pipes, a further allowance has been made for straight couplings at two-metre intervals.

Overheads/profit

Overhead charges are those costs which are not directly linked to any particular contract. They are usually expressed as a percentage of the net cost of labour and materials and are calculated by dividing the total projected annual overhead cost by the projected net turnover. Examples of those items which normally form part of overheads are:

1. head office overheads - heating, lighting, rent, rates, stationery, telephones, secretarial services
2. sundry plant - ladders, towers, vans, etc.
3. insurances - insurance of head office
4. finance charges - interest on overdrafts, bank charges, etc.

The actual percentage used for overheads varies widely. The small plumber working on his own from home who does all his own paperwork and book work in the evening will have few overhead costs and so the percentage addition will be low.

However, the time may come when he finds that he cannot work from home any more and a yard and proper premises must be found, secretary employed, etc., and the percentage will then increase.

In due course these new overheads should be spread over an increased turnover and the percentage will reduce. The profit, of course, is the difference between income and costs and will only be known after the accounts have been prepared. The percentage addition for overheads and profit is 15%.

Total

The total column is the sum of the net plumber labour, net material and overheads/profit columns.

H

CLADDING/COVERING

H71 Lead sheet coverings/flashings

H71 LEAD SHEET COVERINGS/FLASHINGS

	Unit	Material supply (£)	Material waste (%)	Total (£)
MATERIAL COSTS				
Milled sheet lead to BS1178				
code 3	m2	18.64	1.0	18.83
code 4	m2	25.41	1.0	25.66
code 5	m2	31.62	1.0	31.94
code 6	m2	37.48	1.0	37.85
code 7	m2	44.47	1.0	44.91
code 8	m2	50.12	1.0	50.62

	Unit	Labour hours	Net labour (£)	Net material (£)	O'heads /profit (£)	Total (£)
LEAD SHEET COVERINGS/FLASHINGS						
Roof coverings, milled sheet lead to BS1178						
Flat roofing, pitch less than 10 degrees to the horizontal						
code 4	m2	4.00	38.40	25.66	9.61	73.67
code 5	m2	4.20	40.32	31.94	10.84	83.10
code 6	m2	4.40	42.24	37.86	12.02	92.12
code 7	m2	4.60	44.16	44.92	13.36	102.44
code 8	m2	4.80	46.08	50.62	14.50	111.20
Dormers, pitch less than 10 degrees to the horizontal						
code 4	m2	4.75	45.60	25.66	10.69	81.95
code 5	m2	5.00	48.00	31.94	11.99	91.93
code 6	m2	5.25	50.40	37.86	13.24	101.50
code 7	m2	5.25	50.40	44.92	14.30	109.62
code 8	m2	5.75	55.20	50.62	15.87	121.69
Sloping roofing, pitch 10 degrees to 50 degrees						
code 4	m2	4.20	40.32	25.66	9.90	75.88
code 5	m2	4.40	42.24	31.94	11.13	85.31
code 6	m2	4.60	44.16	37.86	12.30	94.32
code 7	m2	4.80	46.08	44.92	13.65	104.65
code 8	m2	5.00	48.00	50.62	14.79	113.41
Dormers, pitch less than 10 degrees to 50 degrees						
code 4	m2	4.75	45.60	25.66	10.69	81.95
code 5	m2	5.00	48.00	31.94	11.99	91.93
code 6	m2	5.25	50.40	37.86	13.24	101.50
code 7	m2	5.50	52.80	44.92	14.66	112.38
code 8	m2	5.75	55.20	50.62	15.87	121.69
Sloping, roofing pitch over 50 degrees						

	Unit	Labour hours	Net labour (£)	Net material (£)	O'heads /profit (£)	Total (£)
Sloping, roofing pitch over 50 degrees						
code 4	m2	4.40	42.24	25.66	10.19	78.09
code 5	m2	4.60	44.16	31.94	11.41	87.51
code 6	m2	4.80	46.08	37.86	12.59	96.53
code 7	m2	5.00	48.00	44.92	13.94	106.86
code 8	m2	5.20	49.92	50.62	15.08	115.62
Dormers, pitch over 50 degrees						
code 4	m2	4.75	45.60	25.66	10.69	81.95
code 5	m2	5.00	48.00	31.94	11.99	91.93
code 6	m2	5.25	50.40	37.86	13.24	101.50
code 7	m2	5.50	52.80	44.92	14.66	112.38
code 8	m2	5.75	55.20	50.62	15.87	121.69
Flashings, horizontal, girth 150mm						
code 4	m	0.40	3.84	3.85	1.15	8.84
code 5	m	0.45	4.32	4.79	1.37	10.48
Flashings, horizontal, girth 200mm						
code 4	m	0.55	5.28	5.13	1.56	11.97
code 5	m	0.60	5.76	6.39	1.82	13.97
Flashings, horizontal, girth 300mm						
code 4	m	0.50	4.80	7.70	1.88	14.38
code 5	m	0.55	5.28	9.58	2.23	17.09
Flashings, sloping, girth 150mm						
code 4	m	0.40	3.84	3.85	1.15	8.84
code 5	m	0.45	4.32	4.79	1.37	10.48
Flashings, sloping, girth 200mm						
code 5	m	0.60	5.76	6.39	1.82	13.97
Flashings, sloping, girth 300mm						
code 4	m	0.80	7.68	7.70	2.31	17.69

Roof coverings (cont'd)	Unit	Labour hours	Net labour (£)	Net material (£)	O'heads /profit (£)	Total (£)
Flashings, sloping, girth 300mm						
code 4	m	0.80	7.68	7.70	2.31	17.69
code 5	m	0.85	8.16	9.58	2.66	20.40
Flashings, stepped, sloping, girth 150mm						
code 4	m	0.50	4.80	3.85	1.30	9.95
code 5	m	0.55	5.28	4.79	1.51	11.58
Flashings, stepped, sloping, girth 200mm						
code 4	m	0.90	8.64	5.13	2.07	15.84
code 5	m	1.00	9.60	6.39	2.40	18.39
Flashings, stepped, sloping, girth 300mm						
code 4	m	1.10	10.56	7.70	2.74	21.00
code 5	m	1.20	11.52	9.58	3.17	24.27
Aprons, horizontal, girth 200mm						
code 4	m	0.55	5.28	5.13	1.56	11.97
code 5	m	1.60	15.36	6.39	3.26	25.01
Aprons, horizontal, girth 300mm						
code 4	m	0.80	7.68	7.70	2.31	17.69
code 5	m	0.90	8.64	9.58	2.73	20.95
Aprons, horizontal, girth 400mm						
code 4	m	1.10	10.56	10.26	3.12	23.94
code 5	m	1.20	11.52	12.78	3.64	27.94
Aprons, sloping, girth 200mm						
code 4	m	0.65	6.24	5.13	1.71	13.08
code 5	m	0.70	6.72	6.39	1.97	15.08
Aprons, sloping, girth 300mm						

	Unit	Labour hours	Net labour (£)	Net material (£)	O'heads /profit (£)	Total (£)
Aprons, sloping, girth 300mm						
code 4	m	0.90	8.64	7.70	2.45	18.79
code 5	m	1.00	9.60	9.58	2.88	22.06
Aprons, sloping, girth 400mm						
code 4	m	1.20	11.52	10.26	3.27	25.05
code 5	m	1.30	12.48	12.78	3.79	29.05
Sills, horizontal, girth 200mm						
code 4	m	0.55	5.28	5.13	1.56	11.97
code 5	m	0.60	5.76	6.39	1.82	13.97
Sills, horizontal, girth 300mm						
code 4	m	0.80	7.68	7.70	2.31	17.69
code 5	m	0.90	8.64	9.58	2.73	20.95
Sills, horizontal, girth 400mm						
code 4	m	1.10	10.56	10.26	3.12	23.94
code 5	m	1.20	11.52	12.78	3.64	27.94
Cappings, horizontal, girth 200mm						
code 4	m	0.55	5.28	5.13	1.56	11.97
code 5	m	0.60	5.76	6.39	1.82	13.97
Cappings, horizontal, girth 300mm						
code 4	m	0.80	7.68	7.70	2.31	17.69
code 5	m	0.90	8.64	9.58	2.73	20.95
Cappings, horizontal, girth 400mm						
code 4	m	1.10	10.56	10.26	3.12	23.94
code 5	m	1.20	11.52	12.78	3.64	27.94
Hips, sloping, girth 200mm						
code 4	m	0.65	6.24	5.13	1.71	13.08
code 5	m	0.75	7.20	6.39	2.04	15.63

Roof coverings (cont'd)	Unit	Labour hours	Net labour (£)	Net material (£)	O'heads /profit (£)	Total (£)
Hips, sloping, girth 300mm						
code 4	m	0.90	8.64	7.70	2.45	18.79
code 5	m	1.00	9.60	9.58	2.88	22.06
Hips, sloping, girth 400mm						
code 4	m	1.20	11.52	10.26	3.27	25.05
code 5	m	1.30	12.48	12.78	3.79	29.05
Kerbs, horizontal, girth 300mm						
code 4	m	0.80	7.68	7.70	2.31	17.69
code 5	m	0.90	8.64	9.58	2.73	20.95
Kerbs, horizontal, girth 400mm						
code 4	m	1.10	10.56	10.26	3.12	23.94
code 5	m	1.20	11.52	12.78	3.64	27.94
Ridges, horizontal, girth 200mm						
code 4	m	0.55	5.28	5.13	1.56	11.97
code 5	m	0.60	5.76	6.39	1.82	13.97
Ridges, horizontal, girth 300mm						
code 4	m	0.80	7.68	7.70	2.31	17.69
code 5	m	0.90	8.64	9.58	2.73	20.95
Ridges, horizontal, girth 400mm						
code 4	m	1.10	10.56	10.26	3.12	23.94
code 5	m	1.20	11.52	12.78	3.64	27.94
Valleys, sloping, girth 400mm						
code 4	m	1.10	10.56	10.26	3.12	23.94
code 5	m	1.20	11.52	12.78	3.64	27.94
Valleys, sloping, girth 600mm						
code 4	m	1.30	12.48	15.40	4.18	32.06
code 5	m	1.40	13.44	19.16	4.89	37.49

CLADDING/COVERING

	Unit	Labour hours	Net labour (£)	Net material (£)	O'heads /profit (£)	Total (£)
Valleys, sloping, girth 800mm						
code 4	m	1.50	14.40	20.53	5.24	40.17
code 5	m	1.60	15.36	25.53	6.13	47.02
Gutters, sloping, girth 400mm						
code 5	m	1.20	11.52	12.78	3.64	27.94
code 6	m	1.40	13.44	15.14	4.29	32.87
code 7	m	1.60	15.36	17.47	4.92	37.75
code 8	m	1.70	16.32	20.25	5.49	42.06
Gutters, sloping, girth 600mm						
code 5	m	1.40	13.44	19.16	4.89	37.49
code 6	m	1.60	15.36	22.72	5.71	43.79
code 7	m	1.80	17.28	26.95	6.63	50.86
code 8	m	1.90	18.24	30.37	7.29	55.90
Gutters, sloping, girth 800mm						
code 5	m	1.60	15.36	25.55	6.14	47.05
code 6	m	1.80	17.28	30.29	7.14	54.71
code 7	m	2.00	19.20	35.94	8.27	63.41
code 8	m	2.10	20.16	40.50	9.10	69.76
Edges, welted						
code 5	m	0.35	3.36	0.00	0.50	3.86
code 6	m	0.45	4.32	0.00	0.65	4.97
code 7	m	0.55	5.28	0.00	0.79	6.07
code 8	m	0.60	5.76	0.00	0.86	6.62
Edges, beaded						
code 5	m	0.35	3.36	0.00	0.50	3.86
code 6	m	0.45	4.32	0.00	0.65	4.97
code 7	m	0.55	5.28	0.00	0.79	6.07
code 8	m	0.60	5.76	0.00	0.86	6.62

Roof coverings (cont'd)	Unit	Labour hours	Net labour (£)	Net material (£)	O'heads /profit (£)	Total (£)
Dressing over corrugated sheeting, down corrugations						
code 5	m	0.20	1.92	0.00	0.29	2.21
code 6	m	0.23	2.21	0.00	0.33	2.54
code 7	m	0.30	2.88	0.00	0.43	3.31
code 8	m	0.40	3.84	0.00	0.58	4.42
Dressing over corrugated sheeting, across corrugations						
code 5	m	1.05	10.08	0.00	1.51	11.59
code 6	m	1.15	11.04	0.00	1.66	12.70
code 7	m	1.20	11.52	0.00	1.73	13.25
code 8	m	1.25	12.00	0.00	1.80	13.80
Dressing over slating and tiling						
code 5	m	0.20	1.92	0.00	0.29	2.21
code 6	m	0.25	2.40	0.00	0.36	2.76
code 7	m	0.30	2.88	0.00	0.43	3.31
code 8	m	0.40	3.84	0.00	0.58	4.42
Dressing over glass and glazing bar						
code 5	m	0.20	1.92	0.00	0.29	2.21
code 6	m	0.25	2.40	0.00	0.36	2.76
code 7	m	0.30	2.88	0.00	0.43	3.31
code 8	m	0.40	3.84	0.00	0.58	4.42
Soakers, size 150 x 150mm						
code 3	nr	0.25	2.40	0.48	0.43	3.31
code 4	nr	0.30	2.88	0.60	0.52	4.00
Soakers, size 200 x 300mm						
code 3	nr	0.30	2.88	1.13	0.60	4.61
code 4	nr	0.35	3.36	1.60	0.74	5.70
Soakers, size 300 x 150mm						
code 3	nr	0.35	3.36	0.85	0.63	4.84
code 4	nr	0.40	3.84	1.20	0.76	5.80

	Unit	Labour hours	Net labour (£)	Net material (£)	O'heads /profit (£)	Total (£)
Slates, size 400 x 400mm with 200mm high collar, 100mm diameter						
code 4	nr	1.50	14.40	5.64	3.01	23.05
code 5	nr	1.60	15.36	7.02	3.36	25.74
Slates, size 400 x 400mm with 200mm high collar, 150mm diameter						
code 4	nr	1.70	16.32	6.41	3.41	26.14
code 5	nr	1.80	17.28	7.99	3.79	29.06
Dots, cast lead						
code 3	nr	0.80	7.68	1.00	1.30	9.98
code 4	nr	0.85	8.16	1.10	1.39	10.65
code 5	nr	0.90	8.64	1.20	1.48	11.32

N

FURNITURE AND EQUIPMENT

	Unit	Material supply (£)	Material waste (%)	Total (£)
MATERIAL COSTS				
BATHS				
Acrylic				
3mm thick, white and coloured	nr	88.20	2.5	90.41
5mm thick, white and coloured	nr	116.40	2.5	119.31
Pressed steel, vitreous enamelled, standard gauge				
white	nr	91.00	2.5	93.28
coloured	nr	100.75	2.5	103.27
Pressed steel vitreous enamelled heavy gauge				
white	nr	245.00	2.5	251.13
coloured	nr	280.00	2.5	287.00
Bath panel, hardboard				
end	nr	2.50	2.5	2.56
side	nr	4.90	2.5	5.02
Bath panel, acrylic				
end	nr	14.60	2.5	14.97
side	nr	22.30	2.5	22.86
Aluminium angle strip	nr	1.75	2.5	1.79
BASINS AND PEDESTALS				
560 x 430mm				
wall-mounted	nr	32.25	2.5	33.06
with pedestal	nr	51.50	2.5	52.79
590 x 440mm				
white medical wall-mounted	nr	61.90	2.5	63.45

	Unit	Material supply (£)	Material waste (%)	Total (£)
N13 SANITARY APPLIANCES				
MATERIAL COSTS				
BATHS				
Acrylic				
3mm thick, white and coloured	nr	88.20	2.5	90.41
5mm thick, white and coloured	nr	116.40	2.5	119.31
Pressed steel, vitreous enamelled, standard gauge				
white	nr	91.00	2.5	93.28
coloured	nr	100.75	2.5	103.27
Pressed steel vitreous enamelled heavy gauge				
white	nr	245.00	2.5	251.13
coloured	nr	280.00	2.5	287.00
Bath panel, hardboard				
end	nr	2.50	2.5	2.56
side	nr	4.90	2.5	5.02
Bath panel, acrylic				
end	nr	14.60	2.5	14.97
side	nr	22.30	2.5	22.86
Aluminium angle strip	nr	1.75	2.5	1.79
BASINS AND PEDESTALS				
560 x 430mm				
wall-mounted	nr	32.25	2.5	33.06
with pedestal	nr	51.50	2.5	52.79
590 x 440mm				

	Unit	Material supply (£)	Material waste (%)	Total (£)
495 x 420mm				
corner wall-mounted white or coloured	nr	31.69	2.5	32.48
550 x 450mm				
vanity basin white or coloured	nr	55.67	2.5	57.06
SINKS				
Stainless steel sink unit				
single bowl, double drainer, 1000 x 500mm	nr	82.00	2.5	84.05
single bowl, double drainer 1500 x 500mm	nr	82.00	2.5	84.05
double bowl, single drainer 1500 x 500mm	nr	249.96	2.5	256.21
Fireclay sink, Belfast pattern				
455 x 380 x 205mm	nr	79.35	2.5	81.33
610 x 455 x 255mm	nr	126.05	2.5	129.20
760 x 455 x 255mm	nr	163.45	2.5	167.54
Fireclay sink, shelf pattern				
610 x 535 x 255mm	nr	145.73	2.5	149.37
WC SUITES				
Close-coupled, white				
washdown	nr	144.80	2.5	148.42
siphonic	nr	263.41	2.5	270.00
Close-coupled, coloured				
washdown	nr	160.86	2.5	164.88
siphonic	nr	267.52	2.5	274.21

MATERIAL COSTS

	Unit	Material supply (£)	Material waste (%)	Total (£)
BIDETS				
Freestanding, plain rim, white	nr	107.80	2.5	110.49
Freestanding, plain rim, coloured	nr	138.90	2.5	142.37
URINALS				
Bowl urinals, white vitreous china				
430 x 305mm	nr	68.65	2.5	70.37
500 x 350mm	nr	131.10	2.5	134.38
Division piece, white				
600 x 300mm	nr	26.40	2.5	27.06
Stainless steel flush pipes				
one-bowl set	nr	18.45	2.5	18.91
two-bowl set	nr	36.90	2.5	37.82
three-bowl set	nr	55.40	2.5	56.79
Automatic flushing cistern, white				
4.5 litre	nr	50.00	2.5	51.25
9 litre	nr	58.85	2.5	60.32
13.5 litre	nr	69.00	2.5	70.73
Slab urinal				
two persons	nr	756.00	2.5	774.90
three persons	nr	980.00	2.5	1004.50
four persons	nr	1205.75	2.5	1235.89
TAPS				
Aztec, chromium-plated				
13mm	nr	32.20	2.5	33.01
19mm	nr	45.15	2.5	46.28

Taps (cont'd)	Unit	Material supply (£)	Material waste (%)	Total (£)
Amazon, chromium-plated				
13mm	nr	11.38	2.5	11.66
19mm	nr	12.60	2.5	12.92
Sink pillar taps				
Standard	nr	20.45	2.5	20.96
Starlite	nr	33.60	2.5	34.44
Mixer taps				
Basin, Streamline	nr	39.80	2.5	40.79
Basin, Streamline	nr	19.60	2.5	20.09
Bath with handspray, Streamline	nr	127.00	2.5	130.18
Bath, Supanova, pillar	nr	77.40	2.5	79.34
Bath with handspray, Starlite	nr	122.55	2.5	125.61
Standard sink with spout	nr	36.00	2.5	36.90
Starlite sink with spout	nr	71.00	2.5	72.78
Sink with monoblock,swivel spout	nr	71.50	2.5	73.29
Sink with hot rinse	nr	71.50	2.5	73.29
Bidet fittings				
Monoblock and spray	nr	118.25	2.5	121.21
WASTE DISPOSAL UNITS				
Tweeny	nr	189.20	2.5	193.93
Carror	nr	110.70	2.5	113.47
VALVES				
Mira valves				
type RADA 15	nr	369.50	2.5	378.74
type RADA 17	nr	438.80	2.5	449.77
type RADA 20	nr	484.50	2.5	496.61
type RADA 25	nr	670.25	2.5	687.01
type 88	nr	86.00	2.5	88.15
type 88B	nr	86.00	2.5	88.15

	Unit	Material supply (£)	Material waste (%)	Total (£)
SHOWER KITS				
Mira				
type SF1/100	nr	53.75	2.5	55.09
type NW SF1/300	nr	53.75	2.5	55.09
type 881	nr	365.50	2.5	374.64
Excel	nr	462.25	2.5	473.81
Heatrae shower systems				
Cameo plus	nr	75.70	1.0	76.46
Carousel standard	nr	124.20	1.0	125.44
Accolade electronic standard	nr	194.50	1.0	196.45
Shower trays				
fireclay	nr	132.25	1.0	133.57
acrylic	nr	49.45	1.0	49.94
'Daryl' shower enclosure				
pivot door, silver	nr	485.00	1.0	489.85
side panel, gold	nr	260.00	1.0	262.60
side panel, silver	nr	187.00	1.0	188.87
'Matki' bathscreen				
NWMV 700, white	nr	223.60	1.0	225.84
'Matki' shower enclosures				
XE75, corner entry, white	nr	641.00	1.0	647.41
NW7375, side panel, silver	nr	251.50	1.0	254.02
NW7375, side panel, gold	nr	268.75	1.0	271.44
CH7578, side panel	nr	192.25	1.0	194.17

	Unit	Labour hours	Net labour (£)	Net material (£)	O'heads /profit (£)	Total (£)
N13 SANITARY APPLIANCES						
BATHS						
Acrylic, reinforced, complete with 2nr chromium-plated clips, waste, overflow with chain and plastic plug, excluding taps and trap						
1700mm long						
white	nr	2.55	24.48	92.67	17.57	134.72
coloured	nr	2.55	24.48	122.29	22.02	168.79
Standard gauge (1.6mm) pressed steel vitreous enamelled, complete with 2nr chromium-plated grips, waste overflow with chain and plastic plug, excluding taps and trap						
1700mm long						
white	nr	2.65	25.44	95.61	18.16	139.21
coloured	nr	2.65	25.44	105.85	19.69	150.98
Heavy gauge (2.3mm) pressed steel vitreous enamelled to BS1344 (no grips) slip-resistant, waste overflow with chain and plastic plug, excluding taps and trap						
1700mm long						
white	nr	3.15	30.24	257.40	43.15	330.79
coloured	nr	3.15	30.24	294.18	48.66	373.08
Bath accessories						
Bath panels, enamelled hardboard, fixing with chromium dome-headed screws, trimming as required						
end panel, BP 2.06	nr	0.30	2.88	2.62	0.82	6.33
side panel, BP 3.92	nr	0.45	4.32	5.15	1.42	10.89

	Unit	Labour hours	Net labour (£)	Net material (£)	O'heads /profit (£)	Total (£)
Bath panels, moulded acrylic, fixing in position for trimming as required						
end panel	nr	0.25	2.40	15.33	2.66	20.39
side panel	nr	0.40	3.84	23.39	4.08	31.31
Angle strip polished aluminium, fixing with chromium-plated dome-headed screws, cutting to length						
25 x 25 x 560mm long	nr	0.30	2.88	1.83	0.71	5.42
BASINS AND PEDESTALS						
Wash basin, vitreous china, complete with chromium-plated waste, overflow with chain and plastic plug, excluding taps and trap						
560 x 430mm, white or coloured, wall-mounted on brackets	nr	1.75	16.80	33.89	7.60	58.29
pedestal-mounted, bedded in mastic, screwed to wall	nr	2.05	19.68	54.08	11.06	84.82
590 x 440mm, white medical, wall-mounted on brackets	nr	1.85	17.76	61.94	11.96	91.66
Corner basin						
495mm wide x 420mm projection, white or coloured wall-mounted on brackets	nr	1.80	17.28	33.29	7.59	58.16
Vanity basin						
550mm x 450mm, white or coloured set into unit	nr	1.70	16.32	58.70	11.25	86.27

Basins and pedestals (cont'd)	Unit	Labour hours	Net labour (£)	Net material (£)	O'heads /profit (£)	Total (£)
Corner basin						
495mm wide x 420mm projection, white or coloured wall-mounted on brackets	nr	1.80	17.28	33.29	7.59	58.16
Vanity basin						
550mm x 450mm, white or coloured set into unit	nr	1.70	16.32	58.70	11.25	86.27
SINKS						
Stainless steel sink unit sinks, complete with inset chromium-plated waste, overflow with chain and plastic plug, excluding taps and trap, fixing to sink top with proprietary clips						
single bowl with single drainer 1000 x 500mm	nr	1.80	17.28	86.15	15.51	118.94
single bowl with double drainer 1500 x 500mm	nr	1.90	18.24	86.15	15.66	120.05
double bowl with single drainer 1500 x 500mm	nr	1.95	18.72	262.62	42.20	323.54
Fireclay sinks, BS1206, white glazed, complete with chromium-plated waste, chain and plastic plug, excluding taps and trap, wall-mounted on brackets						
Belfast pattern						
455 x 380 x 205mm	nr	2.25	21.60	83.36	15.74	120.70
610 x 455 x 255mm	nr	2.35	22.56	132.43	23.25	178.24
760 x 455 x 255mm	nr	2.70	25.92	171.73	29.65	227.30
Shelf pattern						
610 x 535 x 255mm	nr	2.50	24.00	153.10	26.56	203.67

	Unit	Labour hours	Net labour (£)	Net material (£)	O'heads /profit (£)	Total (£)
coloured						
washdown type	nr	2.35	22.56	169.00	28.73	220.29
siphonic type	nr	2.35	22.56	281.07	45.54	349.17
BIDETS						
Freestanding plain rim, vitreous china, excluding fittings						
white	nr	1.95	18.72	113.25	19.80	151.77
coloured	nr	1.95	18.72	145.93	24.70	189.35
URINALS						
Bowl urinals, white vitreous china, wall-mounted on hangers, including chromium-plated dome outlet						
430mm wide x 305mm high	nr	0.90	8.64	72.13	12.12	92.89
500mm wide x 350mm high	nr	0.95	9.12	137.74	22.03	168.89
Division piece, white vitreous china, wall-mounted on hangers						
600mm high x 300mm projection	nr	0.50	4.80	27.74	4.88	37.42
Stainless steel flush pipes and spreaders for bowl urinals, faced fixed to wall						
one-bowl set	nr	1.10	10.56	19.38	4.49	34.43
two-bowl set	nr	1.65	15.84	38.77	8.19	62.80
three-bowl set	nr	2.05	19.68	58.20	11.68	89.56
Automatic flushing cistern and fittings, white vitreous china, wall-mounted on brackets						
4.5 litre	nr	0.95	9.12	52.53	9.25	70.90
9 litre	nr	1.10	10.56	61.83	10.86	83.25
13.5 litre	nr	1.30	12.48	72.49	12.75	97.72

Urinals (cont'd)	Unit	Labour hours	Net labour (£)	Net material (£)	O'heads /profit (£)	Total (£)
Modular slab urinal, white fire-clay china, comprising backslabs without divisions, waterway channel, automatic flushing cistern, stainless steel flush and sparge pipes, fixing in position						
two persons	nr	5.00	48.00	794.27	126.34	968.61
three persons	nr	5.95	57.12	1029.61	163.01	1249.74
four persons	nr	6.90	66.24	1266.79	199.95	1532.98
TAPS						
Basin and bath pillar taps						
Aztec chromium-plated						
13mm	nr	0.35	3.36	33.84	5.58	42.78
19mm	nr	0.35	3.36	47.44	7.62	58.42
Amazon chromium-plated						
13mm	nr	0.35	3.36	11.95	2.30	17.61
19mm	nr	0.35	3.36	13.23	2.49	19.08
Sink pillar taps						
Pegler's chromium-plated	nr	0.35	3.36	21.48	3.73	28.57
Streamline chromium-plated	nr	0.35	3.36	35.51	5.83	44.70
Basin mixer taps						
Streamline chromium-plated, 3 hole pop-up waste with fixed head	nr	0.50	4.80	41.81	6.99	53.60
Swanlux chromium-plated, monoblock pop-up waste with fixed head	nr	0.50	4.80	20.59	3.81	29.20

	Unit	Labour hours	Net labour (£)	Net material (£)	O'heads /profit (£)	Total (£)
Basin mixer taps						
Streamline chromium-plated, 3 hole pop-up waste with fixed head	nr	0.50	4.80	41.81	6.99	53.60
Swanlux chromium-plated, monoblock pop-up waste with fixed head	nr	0.50	4.80	20.59	3.81	29.20
Bath mixer taps						
Streamline chromium-plated, deck pattern with hose and handspray	nr	0.70	6.72	133.43	21.02	161.17
Streamline chromium-plated, deck pattern, filler only	nr	0.55	5.28	81.32	12.99	99.59
Streamline chromium-plated, deck pattern with handspray	nr	0.70	6.72	128.75	20.32	155.79
Sink mixer taps						
Streamline chromium-plated, deck pattern with swivel spout	nr	0.40	3.84	37.82	6.25	47.91
Streamline chromium-plated, deck pattern with swivel spout	nr	0.40	3.84	74.60	11.77	90.21
Streamline studio chromium-plated, monoblock with swivel spout	nr	0.40	3.84	75.12	11.84	90.80
Streamline studio chromium-plated, monoblock with swivel spout, hot rinse attachment	nr	0.50	4.80	75.12	11.99	91.91
Bidet fittings						
Twyford's Aztec chromium-plated, monoblock and spray	nr	0.65	6.24	124.24	19.57	150.05

	Unit	Labour hours	Net labour (£)	Net material (£)	O'heads /profit (£)	Total (£)
WASTE DISPOSAL UNITS						
For metal sinks with 89mm waste outlet, excluding electrical connections						
Tweeny	nr	2.00	19.20	198.78	32.70	250.68
Carrol	nr	1.50	14.40	116.31	19.61	150.32
VALVES						
Mira, theromostatic mixing valves, fixing to wall						
Type RADA15	nr	0.50	4.80	388.21	58.95	451.96
Type RADA17	nr	0.50	4.80	461.00	69.87	535.67
Type 915B	nr	0.50	4.80	509.03	77.07	590.90
Type 915BH	nr	0.50	4.80	704.19	106.35	815.34
Mira, mechanical mixing valves, fixing to wall						
Type 88	nr	0.50	4.80	90.35	14.27	109.42
Type 88B, Sovereign	nr	0.50	4.80	90.35	14.27	109.42
SHOWER KITS						
Mira, fixing to wall						
Type SF1/100 handset with flexible tube, connector and holder	nr	0.50	4.80	56.47	9.19	70.46
Type SF1/300 flexible tube with adjustable handspray, sliding bar, NWSF3	nr	0.75	7.20	56.47	9.55	73.22

	Unit	Labour hours	Net labour (£)	Net material (£)	O'heads /profit (£)	Total (£)
Type Powerpak 881, power shower, with mixer, pump, handset and gel dispenser	nr	1.00	9.60	384.01	59.04	452.65
Type Excel, power shower, with thermostatic mixer pump, handset and gel dispenser	nr	1.10	10.56	485.66	74.43	570.65
ELECTRICAL SHOWER SYSTEMS						
Heatrae Sadia, fixing to wall, excluding electrical connections F7						
Type Cameo - 7kW	nr	0.75	7.20	77.22	12.66	97.08
Type Carousel - 7kW	nr	0.75	7.20	126.69	20.08	153.97
Type Accolade - 7kW	nr	0.75	7.20	198.40	30.84	236.44
SHOWERS						
Trays						
White glazed fireclay, fixing in postion, 760 x 760 x 180mm, white	nr	2.50	24.00	134.91	23.84	182.75
Acrylic, fixing in position, 760 x 760 x 180mm	nr	1.50	14.40	50.44	9.73	74.57
Enclosures						
'Daryl' shower enclosure, fixing ends to walls, safety glass						
pivot door, silver	nr	1.00	9.60	494.75	75.65	580.00
side panel, gold	nr	0.70	6.72	265.23	40.79	312.74
side panel, silver	nr	0.70	6.72	190.76	29.62	227.10
'Matki bathscreen', fixing ends to walls						
NWMV 700, white	nr	1.00	9.60	238.30	37.19	285.09

Shower enclosures (cont'd)	Unit	Labour hours	Net labour (£)	Net material (£)	O'heads /profit (£)	Total (£)
'Matki' shower enclosures, fixing ends to walls, toughened glass (unless otherwise stated)						
XE75, corner entry, white	nr	1.00	9.60	653.88	99.52	763.00
NW7375, side panel, silver	nr	1.00	9.60	256.55	39.92	306.07
NW7375, side panel, gold	nr	1.00	9.60	274.15	42.56	326.31
CH7578, side panel	nr	1.00	9.60	196.11	30.86	236.57

P

BUILDING FABRIC SUNDRIES

P31 Holes, chases, covers, supports for services

P31 HOLES, CHASES, COVERS, SUPPORTS FOR SERVICES

Cutting holes for pipes up to 25mm diameter in

	Unit	Labour hours	Net labour (£)	Net material (£)	O'heads /profit (£)	Total (£)
New brickwork 102.5mm thick						
commons	nr	0.50	4.80	0.00	0.72	5.52
facings	nr	0.55	5.28	0.00	0.79	6.07
engineering Class 'B'	nr	0.75	7.20	0.00	1.08	8.28
engineering Class 'A'	nr	0.95	9.12	0.00	1.37	10.49
New blockwork 100mm thick						
concrete blocks	nr	0.45	4.32	0.00	0.65	4.97
Thermalite blocks	nr	0.35	3.36	0.00	0.50	3.86
aerated concrete blocks	nr	0.32	3.07	0.00	0.46	3.53
New brickwork 215mm thick						
commons	nr	0.95	9.12	0.00	1.37	10.49
facings	nr	1.00	9.60	0.00	1.44	11.04
engineering Class 'B'	nr	1.50	14.40	0.00	2.16	16.56
engineering Class 'A'	nr	1.75	16.80	0.00	2.52	19.32
New blockwork 150mm thick						
concrete blocks	nr	0.55	5.28	0.00	0.79	6.07
Thermalite blocks	nr	0.40	3.84	0.00	0.58	4.42
aerated concrete blocks	nr	0.37	3.55	0.00	0.53	4.08
Extra for making good one side						
common brickwork	nr	0.35	3.36	0.00	0.50	3.86
facing brickwork	nr	0.50	4.80	0.00	0.72	5.52
engineering brickwork	nr	0.55	5.28	0.00	0.79	6.07
concrete blockwork	nr	0.30	2.88	0.00	0.43	3.31
Thermalite blockwork	nr	0.25	2.40	0.00	0.36	2.76
aerated concrete blockwork	nr	0.20	1.92	0.00	0.29	2.21

Cutting holes (cont'd)	Unit	Labour hours	Net labour (£)	Net material (£)	O'heads /profit (£)	Total (£)
Cutting holes for pipes from 25mm up to 50mm diameter in						
New brickwork 102.5mm thick						
commons	nr	0.65	6.24	0.00	0.94	7.18
facings	nr	0.70	6.72	0.00	1.01	7.73
engineering Class 'B'	nr	0.95	9.12	0.00	1.37	10.49
engineering Class 'A'	nr	1.20	11.52	0.00	1.73	13.25
New blockwork 100mm thick						
concrete blocks	nr	0.65	6.24	0.00	0.94	7.18
Thermalite blocks	nr	0.45	4.32	0.00	0.65	4.97
aerated concrete blocks	nr	0.40	3.84	0.00	0.58	4.42
New brickwork 215mm thick						
commons	nr	1.10	10.56	0.00	1.58	12.14
facings	nr	1.25	12.00	0.00	1.80	13.80
engineering Class 'B'	nr	1.75	16.80	0.00	2.52	19.32
engineering Class 'A'	nr	2.00	19.20	0.00	2.88	22.08
New blockwork 150mm thick						
concrete blocks	nr	0.70	6.72	0.00	1.01	7.73
Thermalite blocks	nr	0.50	4.80	0.00	0.72	5.52
aerated concrete blocks	nr	0.45	4.32	0.00	0.65	4.97
(for old brickwork multiply x 2)						
Reinforced concrete 150mm thick	nr	0.75	7.20	0.00	1.08	8.28
Reinforced concrete 250mm thick	nr	1.10	10.56	0.00	1.58	12.14
Cutting and pinning ends of pipe supports bracket to						
Concrete	nr	0.30	2.88	0.00	0.43	3.31
Brickwork	nr	0.25	2.40	0.00	0.36	2.76
Blockwork	nr	0.20	1.92	0.00	0.29	2.21

HOLES, CHASES FOR SERVICE

	Unit	Labour hours	Net labour (£)	Net material (£)	O'heads /profit (£)	Total (£)
Making good around pipe support bracket in						
Facing brickwork	nr	0.15	1.44	0.00	0.22	1.66
Plastered walls	nr	0.10	0.96	0.00	0.14	1.10
Tiled walls	nr	0.20	1.92	0.00	0.29	2.21
Cut opening through external wall in two skins, facing bricks and concrete blocks for						
Balanced flue outlet	nr	1.00	9.60	0.00	1.44	11.04
150mm diameter flue pipe	nr	0.75	7.20	0.00	1.08	8.28
18mm overflow pipe	nr	0.15	1.44	0.00	0.22	1.66
Cutting chases for pipes up to 25mm diameter in						
New brickwork						
commons	m	0.30	2.88	0.00	0.43	3.31
facings	m	0.32	3.07	0.00	0.46	3.53
engineering Class 'B'	m	0.35	3.36	0.00	0.50	3.86
engineering Class 'A'	m	0.40	3.84	0.00	0.58	4.42
New blockwork						
concrete blocks	m	0.30	2.88	0.00	0.43	3.31
Thermalite blocks	m	0.20	1.92	0.00	0.29	2.21
aerated concrete blocks	m	0.15	1.44	0.00	0.22	1.66
Existing brickwork						
commons	m	0.40	3.84	0.00	0.58	4.42
facings	m	0.45	4.32	0.00	0.65	4.97
engineering Class 'B'	m	0.50	4.80	0.00	0.72	5.52
engineering Class 'A'	m	0.60	5.76	0.00	0.86	6.62

Cutting chases (cont'd)	Unit	Labour hours	Net labour (£)	Net material (£)	O'heads /profit (£)	Total (£)
Extra over for						
plastered surfaces	m	0.15	1.44	0.00	0.22	1.66
tiled surfaces	m	0.25	2.40	0.00	0.36	2.76
new concrete floor	m	0.35	3.36	0.00	0.50	3.86
existing granolithic floor	m	0.50	4.80	0.00	0.72	5.52
new concrete soffit	m	0.55	5.28	0.00	0.79	6.07
existing concrete soffit	m	0.65	6.24	0.00	0.94	7.18
Cutting chases for two pipes up to 25mm diameter or one pipe 50mm diameter in						
New brickwork						
commons	m	0.40	3.84	0.00	0.58	4.42
facings	m	0.45	4.32	0.00	0.65	4.97
engineering Class 'B'	nr	0.48	4.61	0.00	0.69	5.30
engineering Class 'A'	m	0.60	5.76	0.00	0.86	6.62
New blockwork						
concrete blocks	m	0.40	3.84	0.00	0.58	4.42
Thermalite blocks	m	0.35	3.36	0.00	0.50	3.86
aerated concrete blocks	m	0.25	2.40	0.00	0.36	2.76
Existing brickwork						
commons	m	0.50	4.80	0.00	0.72	5.52
facings	m	0.55	5.28	0.00	0.79	6.07
engineering Class 'B'	m	0.60	5.76	0.00	0.86	6.62
engineering Class 'A'	m	0.65	6.24	0.00	0.94	7.18
Reinforced concrete						
75mm x 75mm deep	m	0.90	8.64	0.00	1.30	9.94
100mm x 100mm deep	m	1.20	11.52	0.00	1.73	13.25

R

DISPOSAL SYSTEMS

	Unit	Material supply (£)	Material waste (%)	Total (£)
R10 RAINWATER PIPEWORK / GUTTERS				
MATERIAL COSTS				
PIPES				
uPVC Terrain 2100 system				
69mm diameter pipe	m	3.62	5.0	3.80
connector	nr	1.31	2.5	1.34
bend, 92.5 degrees	nr	2.41	2.5	2.47
bend, 112.5 degrees	nr	2.15	2.5	2.20
branch	nr	5.10	2.5	5.23
shoe	nr	1.46	2.5	1.50
access pipe	nr	10.34	2.5	10.60
fitting clip	nr	1.06	2.5	1.09
uPVC Terrain 2200 system				
62mm square pipe	m	4.81	5.0	5.05
connector	nr	1.75	2.5	1.79
bend, 92.5 degrees	nr	2.38	2.5	2.44
branch	nr	5.78	2.5	5.92
shoe	nr	1.75	2.5	1.79
offset	nr	3.61	2.5	3.70
access pipe	nr	14.08	2.5	14.43
change piece, square to round adaptor	nr	2.74	2.5	2.81
fitting clip	nr	0.98	2.5	1.00
uPVC Terrain 2300 large capacity system				
75mm square pipe	m	5.78	5.0	6.07
connector	nr	2.19	2.5	2.24
bend, 112.5 degrees	nr	4.40	2.5	4.51
shoe	nr	2.27	2.5	2.33
change piece	nr	3.30	2.5	3.38
offset	nr	9.75	2.5	9.99
access pipe	nr	16.90	2.5	17.32
fitting clip	nr	2.62	2.5	2.69

Pipes (cont'd)	Unit	Material supply (£)	Material waste (%)	Total (£)
uPVC Terrain 82 system				
82mm diameter pipe	m	5.87	5.0	6.16
shoe	nr	6.99	2.5	7.16
bend, 92.5 degrees	nr	7.12	2.5	7.30
bend, 135 degrees	nr	7.12	2.5	7.30
branch, 92.5 degrees	nr	9.90	2.5	10.15
branch, 135 degrees	nr	9.90	2.5	10.15
spigot socket bend, 92.5 degrees	nr	6.88	2.5	7.05
seal ring adaptor	nr	1.80	2.5	1.85
caulking bush	nr	6.04	2.5	6.19
holderbat plastic coated	nr	3.03	2.5	3.11
uPVC Terrain 110 system				
110mm diameter pipe	m	7.89	5.0	8.28
shoe	nr	8.81	2.5	9.03
bend, 92.5 degrees	nr	9.75	2.5	9.99
bend, 112.5 degrees	nr	9.75	2.5	9.99
bend, 135 degrees	nr	9.75	2.5	9.99
access bend, 92.5 degrees	nr	21.87	2.5	22.42
branch, 92.5 degrees	nr	12.75	2.5	13.07
branch, 135 degrees	nr	12.90	2.5	13.22
access branch, 92.5 degrees	nr	24.98	2.5	25.60
spigot socket bend, 92.5 degrees	nr	8.06	2.5	8.26
seal ring adaptor	nr	1.46	2.5	1.50
access pipe, solvent welded socket	nr	12.17	2.5	12.47
access pipe, spigot and socket	nr	11.96	2.5	12.26
caulking bush	nr	6.04	2.5	6.19
holderbat plastic coated	nr	3.11	2.5	3.19
uPVC Marley circular system				
51mm diameter circular downpipe	m	2.43	5.0	2.55
offset, 76mm	nr	3.92	2.5	4.02
offset, 152mm	nr	4.24	2.5	4.35
offset bends, 67.5 degrees	nr	1.53	2.5	1.57
shoe	nr	1.53	2.5	1.57
pipe clip	nr	0.85	2.5	0.87

MATERIAL COSTS

	Unit	Material supply (£)	Material waste (%)	Total (£)
68mm diameter circular downpipe	m	2.47	5.0	2.59
bend, 45 degrees	nr	3.21	2.5	3.29
bend, 67.5 degrees	nr	3.21	2.5	3.29
bend, 87.5 degrees	nr	3.21	2.5	3.29
offset bend, 67.5 degrees	nr	1.78	2.5	1.82
offset bend, 20 degrees	nr	1.93	2.5	1.98
branch, 67.5 degrees	nr	6.40	2.5	6.56
access pipe	nr	8.80	2.5	9.02
reducer, 110 to 68mm	nr	2.77	2.5	2.84
reducer, 82 to 68mm	nr	2.26	2.5	2.32
shoe	nr	2.77	2.5	2.84
clip	nr	0.74	2.5	0.76
one piece pipe clip	nr	0.90	2.5	0.92
socket clip	nr	0.74	2.5	0.76
65mm square downpipe	m	2.89	5.0	3.03
bend, 67.5 degrees	nr	2.03	2.5	2.08
bend, 87.5 degrees	nr	2.33	2.5	2.39
offset, 54mm	nr	3.55	2.5	3.64
branch, 67.5 degrees	nr	7.82	2.5	8.02
access pipe	nr	10.84	2.5	11.11
shoe	nr	2.32	2.5	2.38
drain adaptor	nr	3.40	2.5	3.49
one piece pipe clip	nr	1.08	0.0	1.08
82mm diameter circular downpipe	m	5.30	5.0	5.56
bend, 87.5 degrees	nr	7.02	2.5	7.20
bend, adjustable	nr	11.77	2.5	12.06
offset bend, socket 67.5 degrees	nr	7.02	2.5	7.20
offset bend, spigot 67.5 degrees	nr	5.93	2.5	6.08
branch, adjustable	nr	11.01	2.5	11.29
access pipe	nr	13.50	2.5	13.84
shoe	nr	7.88	2.5	8.08
adaptor	nr	4.55	2.5	4.66
pipe clip	nr	1.65	2.5	1.69
socket clip	nr	3.12	0.0	3.12
110mm diameter circular downpipe	m	5.22	5.0	5.48
bend, 87.5 degrees	nr	9.28	2.5	9.51
bend, adjustable	nr	11.77	2.5	12.06

Pipes (cont'd)	Unit	Material supply (£)	Material waste (%)	Total (£)
offset bend, 67.5 degrees	nr	7.80	2.5	8.00
branch, 87.5 degrees	nr	13.04	2.5	13.37
access pipe, straight	nr	14.76	2.5	15.13
shoe	nr	10.19	2.5	10.44
adaptor	nr	3.81	2.5	3.91
pipe clip	nr	1.58	2.5	1.62
socket clip	nr	3.20	2.5	3.28
160mm diameter circular downpipe	m	13.30	5.0	13.97
bend, adjustable	nr	23.25	2.5	23.83
offset bend, socket 67.5, degrees	nr	21.81	2.5	22.36
offset bend, spigot 67.5, degrees	nr	20.91	2.5	21.43
branch, 87.5 degrees	nr	36.52	2.5	37.43
access pipe	nr	27.21	2.5	27.89
shoe	nr	21.71	2.5	22.25
adaptor	nr	12.17	2.5	12.47
socket clip	nr	4.83	2.5	4.95

uPVC, Osma Minifit system

55mm diameter pipe	m	1.71	5.0	1.80
bend, 45 degrees	nr	1.85	2.5	1.90
bend, 87.5 degrees	nr	1.85	2.5	1.90
offset bend	nr	2.14	2.5	2.19
shoe	nr	1.38	2.5	1.41
pipe bracket	nr	0.82	2.5	0.84

uPVC, Osma Roundline system

68mm diameter pipe	m	2.04	5.0	2.14
bend, 87.5 degrees	nr	2.90	2.5	2.97
offset bend	nr	1.62	2.5	1.66
shoe	nr	1.71	2.5	1.75
branch, 67.5 degrees	nr	5.15	2.5	5.28
access pipe	nr	7.07	2.5	7.25
universal connector	nr	2.74	2.5	2.81
pipe bracket	nr	1.06	2.5	1.09

	Unit	Material supply (£)	Material waste (%)	Total (£)
uPVC, Osma Squareline system				
61mm square pipe	m	2.21	5.0	2.32
bend, 87.5 degrees	nr	2.95	2.5	3.02
offset bend	nr	1.70	2.5	1.74
shoe	nr	2.15	2.5	2.20
branch, 67.5 degrees	nr	5.21	2.5	5.34
access pipe	nr	7.17	2.5	7.35
universal connector	nr	2.74	2.5	2.81
pipe bracket	nr	0.85	2.5	0.87
uPVC, Osma Amazon system				
70mm square pipe	m	2.49	5.0	2.61
offset bend	nr	1.86	2.5	1.91
shoe	nr	2.37	2.5	2.43
branch, 67.5 degrees	nr	5.96	2.5	6.11
access pipe	nr	8.44	2.5	8.65
offset	nr	2.81	2.5	2.88
offset with access	nr	8.44	2.5	8.65
pipe bracket	nr	0.94	2.5	0.96
uPVC, Hepworth system				
50mm diameter pipe	m	2.14	5.0	2.25
offset bend, 112.5 degrees	nr	2.10	2.5	2.15
shoe	nr	2.06	2.5	2.11
saddle bracket	nr	1.30	2.5	1.33
68mm diameter pipe	m	2.24	5.0	2.35
offset bend, 92.5 degrees	nr	3.09	2.5	3.17
offset bend, 112.5 degrees	nr	1.57	2.5	1.61
branch, 112.5 degrees	nr	5.61	2.5	5.75
shoe	nr	1.57	2.5	1.61
access pipe	nr	7.36	2.5	7.54
saddle bracket	nr	0.87	2.5	0.89

Pipes (cont'd)	Unit	Material supply (£)	Material waste (%)	Total (£)
65mm square pipe	m	2.02	5.0	2.12
shoe	nr	2.39	2.5	2.45
offset bend, 112.5 degrees	nr	1.77	2.5	1.81
mini offset, 44mm	nr	4.30	2.5	4.41
adaptor, square to round	nr	2.64	2.5	2.71
branch, 112.5 degrees	nr	5.70	2.5	5.84
access pipe	nr	7.02	2.5	7.20
barrel clip	nr	1.02	2.5	1.05
Aluminium, Alumasc				
63.5mm diameter pipe	m	8.80	5.0	9.24
offset, 76mm	nr	10.70	2.5	10.97
offset, 114mm	nr	11.79	2.5	12.08
offset, 152mm	nr	13.27	2.5	13.60
offset, 229mm	nr	15.05	2.5	15.43
offset, 305mm	nr	16.84	2.5	17.26
offset, 381mm	nr	18.67	2.5	19.14
offset, 457mm	nr	20.45	2.5	20.96
offset, 533mm	nr	22.26	2.5	22.82
offset, 610mm	nr	24.08	2.5	24.68
offset, 686mm	nr	25.83	2.5	26.48
offset, 762mm	nr	27.57	2.5	28.26
two-part offset, 228mm	nr	12.05	2.5	12.35
two-part offset, 381mm	nr	12.98	2.5	13.30
two-part offset, 457mm	nr	13.45	2.5	13.79
two-part offset, 685mm	nr	14.98	2.5	15.35
two-part offset, 914mm	nr	16.23	2.5	16.64
shoe	nr	4.85	2.5	4.97
branch, 92.5 degrees	nr	7.29	2.5	7.47
branch, 112.5 degrees	nr	7.29	2.5	7.47
bend, 92.5 degrees	nr	5.56	2.5	5.70
bend, 112.5 degrees	nr	5.56	2.5	5.70
bend, 135 degrees	nr	5.56	2.5	5.70
pipe clip	nr	2.17	2.5	2.22
76.5mm diameter pipe	m	10.24	5.0	10.75
offset, 76mm	nr	12.97	2.5	13.29
offset, 114mm	nr	13.54	2.5	13.88
offset, 152mm	nr	14.91	2.5	15.28
offset, 229mm	nr	16.64	2.5	17.06

	Unit	Material supply (£)	Material waste (%)	Total (£)
offset, 305mm	nr	18.47	2.5	18.93
offset, 381mm	nr	20.27	2.5	20.78
offset, 457mm	nr	23.70	2.5	24.29
offset, 533mm	nr	25.61	2.5	26.25
offset, 610mm	nr	27.58	2.5	28.27
offset, 686mm	nr	27.58	2.5	28.27
offset, 762mm	nr	31.43	2.5	32.22
two-part offset, 228mm	nr	13.02	2.5	13.35
two-part offset, 381mm	nr	14.14	2.5	14.49
two-part offset, 457mm	nr	14.75	2.5	15.12
two-part offset, 685mm	nr	16.64	2.5	17.06
two-part offset, 914mm	nr	18.26	2.5	18.72
shoe	nr	6.74	2.5	6.91
branch, 92.5 degrees	nr	9.06	2.5	9.29
branch, 112.5 degrees	nr	9.06	2.5	9.29
bend, 92.5 degrees	nr	7.29	2.5	7.47
bend, 112.5 degrees	nr	7.29	2.5	7.47
bend, 135 degrees	nr	16.58	2.5	16.99
pipe clip	nr	2.45	2.5	2.51
102mm diameter pipe	m	17.86	5.0	18.75
offset, 76mm	nr	14.67	2.5	15.04
offset, 114mm	nr	15.80	2.5	16.20
offset, 152mm	nr	17.04	2.5	17.47
offset, 229mm	nr	19.28	2.5	19.76
offset, 305mm	nr	21.54	2.5	22.08
offset, 381mm	nr	23.77	2.5	24.36
offset, 457mm	nr	27.21	2.5	27.89
offset, 533mm	nr	30.02	2.5	30.77
offset, 610mm	nr	31.93	2.5	32.73
offset, 686mm	nr	34.36	2.5	35.22
offset, 762mm	nr	36.61	2.5	37.53
shoe	nr	8.05	2.5	8.25
branch, 92.5 degrees	nr	12.62	2.5	12.94
branch, 112.5 degrees	nr	10.65	2.5	10.92
bend, 92.5 degrees	nr	10.65	2.5	10.92
bend, 112.5 degrees	nr	10.65	2.5	10.92
bend, 135 degrees	nr	20.19	2.5	20.69
pipe clip	nr	2.76	2.5	2.83

Pipes (cont'd)	Unit	Material supply (£)	Material waste (%)	Total (£)
72mm square pipe	m	24.03	5.0	25.23
offset, 76mm	nr	25.37	2.5	26.00
offset, 304mm	nr	34.09	2.5	34.94
offset, 533mm	nr	36.61	2.5	37.53
offset, 762mm	nr	39.07	2.5	40.05
shoe	nr	18.96	2.5	19.43
bend, 112.5 degrees	nr	20.42	2.5	20.93
rodding eye	nr	29.00	2.5	29.73
pipe clip	nr	2.17	2.5	2.22
102 x 76mm rectangular pipe	m	22.14	5.0	23.25
offset, 76mm	nr	29.14	2.5	29.87
offset, 305mm	nr	38.58	2.5	39.54
offset, 533mm	nr	41.80	2.5	42.84
offset, 762mm	nr	44.85	2.5	45.97
shoe	nr	20.42	2.5	20.93
bend, 112.5 degrees	nr	23.34	2.5	23.92
rodding eye	nr	30.63	2.5	31.40
pipe clip	nr	2.45	2.5	2.51
102mm square pipe	m	26.72	5.0	28.06
offset, 76mm	nr	33.83	2.5	34.68
offset, 305mm	nr	44.94	2.5	46.06
offset, 533mm	nr	50.09	2.5	51.34
offset, 762mm	nr	54.71	2.5	56.08
shoe	nr	22.89	2.5	23.46
bend, 112.5 degrees	nr	27.49	2.5	28.18
rodding eye	nr	34.55	2.5	35.41
pipe clip	nr	2.76	2.5	2.83
Cast iron straight, ears cast on, spigot and socket				
50mm diameter pipe	m	12.83	5.0	13.47
bend	nr	6.96	2.5	7.13
offset, 75mm	nr	10.62	2.5	10.89
offset, 230mm	nr	10.62	2.5	10.89
shoe	nr	11.35	2.5	11.63
wall bracket	nr	4.17	2.5	4.27

MATERIAL COSTS

	Unit	Material supply (£)	Material waste (%)	Total (£)
65mm diameter pipe	m	12.83	5.0	13.47
bend	nr	6.96	2.5	7.13
offset, 75mm	nr	10.62	2.5	10.89
offset, 230mm	nr	10.62	2.5	10.89
shoe	nr	11.35	2.5	11.63
wall bracket	nr	4.45	2.5	4.56
75mm diameter pipe	m	13.47	5.0	14.14
bend	nr	8.05	2.5	8.25
offset, 75mm	nr	10.62	2.5	10.89
offset, 230mm	nr	10.62	2.5	10.89
shoe	nr	11.35	2.5	11.63
wall bracket	nr	4.45	2.5	4.56
100mm diameter pipe	m	18.10	5.0	19.01
bend	nr	11.36	2.5	11.64
offset, 75mm	nr	18.19	2.5	18.64
offset, 230mm	nr	10.62	2.5	10.89
shoe	nr	14.75	2.5	15.12
wall bracket	nr	4.45	2.5	4.56

GUTTERS

uPVC, Terrain 2100 system

	Unit	Material supply (£)	Material waste (%)	Total (£)
110mm gutter	m	2.88	5.0	3.02
running outlet	nr	3.07	2.5	3.15
angle, 90 degrees	nr	3.10	2.5	3.18
angle, 120 degrees	nr	3.67	2.5	3.76
angle, 135 degrees	nr	3.20	2.5	3.28
short stop end	nr	1.51	2.5	1.55
stop end outlet	nr	2.96	2.5	3.03
uPVC adaptor to cast iron half-round	nr	4.48	2.5	4.59
aluminium adaptor to cast iron half-round	nr	17.56	2.5	18.00
joint bracket	nr	2.03	2.5	2.08
support bracket	nr	0.78	2.5	0.80

Gutters (cont'd)	Unit	Material supply (£)	Material waste (%)	Total (£)
uPVC, Terrain 2200 square system				
120mm gutter	m	2.94	5.0	3.09
running outlet	nr	3.32	2.5	3.40
angle, 90 degrees	nr	3.10	2.5	3.18
angle, 135 degrees	nr	3.40	2.5	3.49
short stop end	nr	1.66	2.5	1.70
stop end outlet	nr	3.37	2.5	3.45
aluminium adaptor to cast iron ogee	nr	16.60	2.5	17.02
joint bracket	nr	2.02	0.0	2.02
support bracket	nr	0.95	0.0	0.95
uPVC, Terrain 2300 large capacity system				
132mm gutter	m	6.19	5.0	6.50
running outlet	nr	9.26	2.5	9.49
angle, 90 degrees	nr	8.60	2.5	8.82
angle, 135 degrees	nr	16.84	2.5	17.26
short stop end	nr	3.20	2.5	3.28
screwed outlet adaptor	nr	4.87	2.5	4.99
joint bracket	nr	5.27	2.5	5.40
support bracket	nr	2.19	2.5	2.24
uPVC, Marley, Miniline system				
75mm half-round gutter	m	1.38	5.0	1.45
angle, 45 degrees	nr	2.46	2.5	2.52
angle, 90 degrees	nr	2.66	2.5	2.73
outlet	nr	2.07	2.5	2.12
stop end, internal	nr	0.90	2.5	0.92
stop end, external	nr	0.90	2.5	0.92
union clip	nr	0.90	2.5	0.92
fascia bracket	nr	0.54	2.5	0.55

	Unit	Material supply (£)	Material waste (%)	Total (£)
uPVC, Marley, Industrial system				
150mm half-round gutter	m	5.98	5.0	6.28
angle, 90 degrees	nr	11.56	2.5	11.85
outlet	nr	9.78	2.5	10.02
stop end, internal	nr	2.24	2.5	2.30
stop end, external	nr	3.15	2.5	3.23
union clip	nr	4.78	2.5	4.90
fascia bracket	nr	2.59	2.5	2.65
uPVC, Marley, Deepflow system				
110mm semi-elliptical	m	3.22	5.0	3.38
angle, 45 degrees	nr	5.66	2.5	5.80
angle, 90 degrees	nr	3.77	2.5	3.86
outlet	nr	3.45	2.5	3.54
stop end	nr	1.85	2.5	1.90
stop end outlet	nr	3.03	2.5	3.11
union clip	nr	3.08	2.5	3.16
fascia bracket	nr	0.77	2.5	0.79
uPVC, Marley, Premier system				
112mm half-round gutter	m	2.79	5.0	2.93
angle, 45 degrees	nr	4.09	2.5	4.19
angle, 90 degrees	nr	3.89	2.5	3.99
outlet	nr	3.56	2.5	3.65
stop end	nr	1.55	2.5	1.59
stop end outlet	nr	3.56	2.5	3.65
adaptor to ogee	nr	9.18	2.5	9.41
union clip	nr	2.38	2.5	2.44
fascia bracket	nr	0.90	2.5	0.92
uPVC, Marley, Flowline system				
116 x 60mm rectangular gutter	m	2.18	5.0	2.29
angle, 45 degrees	nr	4.91	2.5	5.03
angle, 90 degrees	nr	3.36	2.5	3.44
outlet	nr	3.36	2.5	3.44

Gutters (cont'd)	Unit	Material supply (£)	Material waste (%)	Total (£)
stop end, external	nr	1.49	2.5	1.53
union bracket	nr	1.86	2.5	1.91
fascia bracket	nr	0.82	2.5	0.84
uPVC, Marley, Classic system				
120 x 75mm ogee gutter	m	4.06	2.5	4.16
angle, 45 degrees	nr	7.79	2.5	7.98
angle, 90 degrees	nr	7.44	2.5	7.63
outlet	nr	6.10	2.5	6.25
stop end	nr	1.60	2.5	1.64
union clip	nr	2.70	2.5	2.77
fascia bracket	nr	1.02	2.5	1.05
uPVC, Osma Minifit system				
75mm half-round gutter	m	1.27	5.0	1.33
running outlet	nr	1.80	2.5	1.85
angle, 45 degrees	nr	1.98	2.5	2.03
angle, 90 degrees	nr	1.76	2.5	1.80
stop end	nr	0.78	2.5	0.80
stop end outlet	nr	1.76	2.5	1.80
jointing bracket	nr	0.93	2.5	0.95
support bracket	nr	0.49	2.5	0.50
uPVC, Osma Roundline system, joint bracket joints, to timber with support brackets at 1m maximum centres				
112mm half-round gutter	m	2.04	5.0	2.14
swivelock running outlet	nr	2.84	2.5	2.91
swivelock running outlet with offset bend	nr	3.95	2.5	4.05
angle, 45 degrees	nr	2.93	2.5	3.00
angle, 60 degrees	nr	2.93	2.5	3.00
angle, 90 degrees	nr	2.66	2.5	2.73
stop end, internal	nr	0.71	2.5	0.73
stop end, external	nr	1.14	2.5	1.17
connector to half-round cast iron gutter	nr	2.50	2.5	2.56

	Unit	Material supply (£)	Material waste (%)	Total (£)
connector to ogee cast iron gutter	nr	3.96	2.5	4.06
jointing bracket	nr	1.37	2.5	1.40
support bracket	nr	0.72	2.5	0.74
uPVC, Osma Superline system				
125mm half-round gutter	m	2.72	5.0	2.86
running outlet	nr	3.00	2.5	3.08
stop end, internal	nr	0.96	2.5	0.98
stop end, external	nr	1.47	2.5	1.51
angle, 45 degrees	nr	3.51	2.5	3.60
angle, 90 degrees	nr	3.34	2.5	3.42
gutter union	nr	2.31	2.5	2.37
support bracket	nr	0.94	2.5	0.96
uPVC, Osma Roofline system				
150mm half-round gutter	m	5.56	5.0	5.84
running outlet	nr	7.53	2.5	7.72
stop end, internal	nr	1.19	2.5	1.22
stop end, external	nr	3.74	2.5	3.83
angle, 90 degrees	nr	9.77	2.5	10.01
gutter union	nr	5.48	2.5	5.62
support bracket	nr	2.21	2.5	2.27
uPVC, Osma Amazon system				
110 x 75mm rectangular gutter	m	2.69	5.0	2.82
running outlet	nr	3.19	2.5	3.27
stop end, internal	nr	1.22	2.5	1.25
stop end, external	nr	1.71	2.5	1.75
angle, 45 degrees	nr	3.36	2.5	3.44
angle, 90 degrees	nr	3.20	2.5	3.28
gutter union	nr	2.50	2.5	2.56
support bracket	nr	0.93	2.5	0.95

Gutters (cont'd)	Unit	Material supply (£)	Material waste (%)	Total (£)
uPVC, Osma Squareline system				
100mm gutter	m	2.08	5.0	2.18
running outlet	nr	2.90	2.5	2.97
stop end, internal	nr	0.91	2.5	0.93
stop end, external	nr	1.29	2.5	1.32
angle, 45 degrees	nr	2.96	2.5	3.03
angle, 90 degrees	nr	2.81	2.5	2.88
gutter union	nr	2.29	2.5	2.35
support bracket	nr	0.76	2.5	0.78
uPVC, Hepworth system				
76mm half-round gutter	m	1.57	5.0	1.65
stop end outlet	nr	2.42	2.5	2.48
running outlet	nr	2.42	2.5	2.48
stop end	nr	1.04	2.5	1.07
angle, 90 degrees	nr	2.44	2.5	2.50
floating union	nr	1.18	2.5	1.21
support bracket	nr	0.72	2.5	0.74
112mm half-round gutter	m	2.17	5.0	2.28
running outlet	nr	2.12	2.5	2.17
stop end	nr	1.18	2.5	1.21
angle, 90 degrees	nr	2.44	2.5	2.50
angle, 120 degrees	nr	2.80	2.5	2.87
angle, 135 degrees	nr	2.80	2.5	2.87
adaptor to metal ogee	nr	4.96	2.5	5.08
floating union	nr	1.50	2.5	1.54
support bracket	nr	0.72	0.0	0.72
150mm half-round gutter	m	5.85	5.0	6.14
stop end outlet	nr	7.38	2.5	7.56
running outlet	nr	7.68	2.5	7.87
stop end	nr	3.51	2.5	3.60
angle, 90 degrees	nr	10.13	2.5	10.38
support bracket	nr	2.41	2.5	2.47

	Unit	Material supply (£)	Material waste (%)	Total (£)
112mm square gutter	m	2.25	5.0	2.36
running outlet	nr	3.10	2.5	3.18
stop end	nr	1.10	2.5	1.13
angle, 90 degrees	nr	3.00	2.5	3.08
angle, 120 degrees	nr	3.80	2.5	3.90
angle, 135 degrees	nr	3.80	2.5	3.90
floating union	nr	1.91	2.5	1.96
support bracket	nr	1.77	2.5	1.81
115 x 75mm high-capacity gutter	m	2.68	5.0	2.81
running outlet	nr	3.59	2.5	3.68
stop end	nr	1.71	2.5	1.75
angle, 90 degrees	nr	3.48	2.5	3.57
angle, 135 degrees	nr	5.18	2.5	5.31
adaptor to other gutters	nr	2.87	2.5	2.94
adaptor to square down pipe	nr	2.54	2.5	2.60
support bracket	nr	0.66	2.5	0.68
Aluminium, Alumasc system				
100mm plain half-round gutter	m	6.00	5.0	6.30
angle, 90 degrees	nr	3.67	2.5	3.76
angle, 135 degrees	nr	3.67	2.5	3.76
outlet	nr	4.23	2.5	4.34
stop end outlet	nr	3.49	2.5	3.58
stop end	nr	1.78	2.5	1.82
fascia bracket	nr	1.04	2.5	1.07
100mm, plain half-round, polyester powder-coated gutter	m	10.22	5.0	10.73
angle, 90 degrees	nr	5.65	2.5	5.79
angle, 135 degrees	nr	5.65	2.5	5.79
outlet	nr	6.12	2.5	6.27
stop end outlet	nr	5.45	2.5	5.59
stop end	nr	2.76	2.5	2.83
fascia bracket	nr	1.46	2.5	1.50

Gutters (cont'd)	Unit	Material supply (£)	Material waste (%)	Total (£)
113mm plain, half-round gutter	m	6.36	5.0	6.68
angle, 90 degrees	nr	4.20	2.5	4.31
angle, 135 degrees	nr	4.20	2.5	4.31
outlet	nr	4.52	2.5	4.63
stop end outlet	nr	3.91	2.5	4.01
stop end	nr	1.82	2.5	1.87
fascia bracket	nr	1.06	2.5	1.09
113mm plain half-round, polyester powder-coated gutter	m	10.72	5.0	11.26
angle, 90 degrees	nr	6.38	2.5	6.54
angle, 135 degrees	nr	6.38	2.5	6.54
outlet	nr	6.67	2.5	6.84
stop end outlet	nr	6.25	2.5	6.41
stop end	nr	2.89	2.5	2.96
fascia bracket	nr	1.49	2.5	1.53
125mm plain, half-round gutter	m	7.50	5.0	7.88
angle, 90 degrees	nr	4.70	2.5	4.82
angle, 135 degrees	nr	4.70	2.5	4.82
outlet	nr	4.84	2.5	4.96
stop end	nr	2.36	2.5	2.42
fascia bracket	nr	1.92	2.5	1.97
125mm plain, half-round, polyester powder-coated gutter	m	12.03	5.0	12.63
angle, 90 degrees	nr	7.09	2.5	7.27
angle, 135 degrees	nr	7.09	2.5	7.27
outlet	nr	7.23	2.5	7.41
stop end	nr	3.53	2.5	3.62
fascia bracket	nr	2.73	2.5	2.80
113mm, beaded, plain, half-round gutter	m	7.73	5.0	8.12
angle, 90 degrees	nr	5.86	2.5	6.01
outlet	nr	6.17	2.5	6.32
stop end	nr	3.26	2.5	3.34
fascia bracket	nr	1.82	2.5	1.87

MATERIAL COSTS

	Unit	Material supply (£)	Material waste (%)	Total (£)
113mm, beaded, half-round, polyester powder-coated gutter	m	13.04	5.0	13.69
angle, 90 degrees	nr	8.19	2.5	8.39
outlet	nr	8.50	2.5	8.71
stop end	nr	4.38	2.5	4.49
fascia bracket	nr	2.89	2.5	2.96
125mm, beaded, plain, half-round gutter	m	9.13	5.0	9.59
angle, 90 degrees	nr	6.47	2.5	6.63
outlet	nr	6.62	2.5	6.79
stop end	nr	3.81	2.5	3.91
fascia bracket	nr	2.36	2.5	2.42
125mm, beaded, half-round, polyester powder-coated	m	14.65	5.0	15.38
angle, 90 degrees	nr	9.06	2.5	9.29
outlet	nr	9.16	2.5	9.39
stop end	nr	5.03	2.5	5.16
fascia bracket	nr	3.53	2.5	3.62
100mm, Victorian ogee, plain gutter	m	7.90	5.0	8.29
angle, 90 degrees	nr	4.08	2.5	4.18
angle, 135 degrees	nr	4.08	2.5	4.18
stop end outlet	nr	3.64	2.5	3.73
stop end	nr	1.94	2.5	1.99
fascia bracket	nr	1.94	2.5	1.99
100mm, Victorian ogee, polyester powder-coated gutter	m	12.75	5.0	13.39
angle, 90 degrees	nr	6.05	2.5	6.20
angle, 135 degrees	nr	6.05	2.5	6.20
stop end outlet	nr	5.57	2.5	5.71
stop end	nr	2.90	2.5	2.97
fascia bracket	nr	2.90	2.5	2.97

Gutters (cont'd)	Unit	Material supply (£)	Material waste (%)	Total (£)
113mm, Victorian ogee, plain gutter	m	9.15	5.0	9.61
angle, 90 degrees	nr	5.05	2.5	5.18
angle, 135 degrees	nr	5.05	2.5	5.18
stop end outlet	nr	4.07	2.5	4.17
stop end	nr	2.05	2.5	2.10
fascia bracket	nr	2.05	2.5	2.10
113mm, Victorian ogee, polyester powder-coated gutter	m	14.19	5.0	14.90
angle, 90 degrees	nr	7.23	2.5	7.41
angle, 135 degrees	nr	7.23	2.5	7.41
stop end outlet	nr	6.24	2.5	6.40
stop end	nr	3.12	2.5	3.20
fascia bracket	nr	3.12	2.5	3.20
125mm, Victorian ogee, plain gutter	m	10.43	5.0	10.95
angle, 90 degrees	nr	6.04	2.5	6.19
angle, 135 degrees	nr	6.04	2.5	6.19
stop end outlet	nr	4.70	2.5	4.82
stop end	nr	2.21	2.5	2.27
fascia bracket	nr	2.21	2.5	2.27
125mm, Victorian ogee, polyester powder-coated gutter	m	15.67	5.0	16.45
angle, 90 degrees	nr	8.43	2.5	8.64
angle, 135 degrees	nr	8.43	2.5	8.64
stop end outlet	nr	7.09	2.5	7.27
stop end	nr	3.40	2.5	3.48
fascia bracket	nr	3.40	2.5	3.48
Moulded No. 46, 100 x 75mm, plain gutter	m	11.83	5.0	12.42
angle, 90 degrees	nr	4.62	2.5	4.74
angle, 135 degrees	nr	4.62	2.5	4.74
outlet for 63mm pipe	nr	4.77	2.5	4.89

	Unit	Material supply (£)	Material waste (%)	Total (£)
outlet for 76mm pipe	nr	14.05	2.5	14.40
outlet for 102 x 76mm pipe	nr	23.51	2.5	24.10
fascia bracket	nr	3.17	2.5	3.25
Moulded No. 46, 100 x 75mm, polyester powder-coated gutter	m	17.06	5.0	17.91
angle, 90 degrees	nr	6.83	2.5	7.00
angle, 135 degrees	nr	6.83	2.5	7.00
outlet for 63mm pipe	nr	6.83	2.5	7.00
outlet for 76mm pipe	nr	16.22	2.5	16.63
outlet for 102 x 76mm pipe	nr	26.37	2.5	27.03
fascia bracket	nr	4.23	2.5	4.34
Moulded No. 46, 125 x 100mm, plain gutter	m	15.23	5.0	15.99
angle, 90 degrees	nr	7.42	2.5	7.61
angle, 135 degrees	nr	7.42	2.5	7.61
outlet for 63mm pipe	nr	7.42	2.5	7.61
outlet for 76mm pipe	nr	7.42	2.5	7.61
outlet for 102 x 76mm pipe	nr	7.42	2.5	7.61
outlet for 102 x 102mm pipe	nr	28.46	2.5	29.17
fascia bracket	nr	3.83	2.5	3.93
Moulded No. 46, 125 x 100mm, polyester powder-coated gutter	m	20.74	5.0	21.78
angle, 90 degrees	nr	9.83	2.5	10.08
angle, 135 degrees	nr	9.83	2.5	10.08
outlet for 63mm pipe	nr	9.83	2.5	10.08
outlet for 76mm pipe	nr	9.83	2.5	10.08
outlet for 102 x 76mm pipe	nr	9.83	2.5	10.08
outlet for 102 x 102mm pipe	nr	31.32	2.5	32.10
fascia bracket	nr	5.00	2.5	5.13
Moulded No. 46, 150 x 100mm, plain gutter	m	19.73	5.0	20.72
angle, 90 degrees	nr	17.17	2.5	17.60
angle, 135 degrees	nr	17.17	2.5	17.60
outlet for 63mm pipe	nr	17.17	2.5	17.60
outlet for 76mm pipe	nr	17.17	2.5	17.60

Gutters (cont'd)	Unit	Material supply (£)	Material waste (%)	Total (£)
outlet for 102 x 76mm pipe	nr	32.91	2.5	33.73
outlet for 102 x 102mm pipe	nr	32.91	2.5	33.73
fascia bracket	nr	4.71	2.5	4.83
Moulded No. 46, 150 x 100mm, polyester powder-coated gutter	m	25.82	5.0	27.11
angle, 90 degrees	nr	19.78	2.5	20.27
angle, 135 degrees	nr	19.78	2.5	20.27
outlet for 63mm pipe	nr	19.78	2.5	20.27
outlet for 76mm pipe	nr	19.78	2.5	20.27
outlet for 102 x 76mm pipe	nr	35.51	2.5	36.40
outlet for 102 x 102mm pipe	nr	35.51	2.5	36.40
fascia bracket	nr	6.93	2.5	7.10
Beaded deep run, 113 x 75mm, plain gutter	m	12.91	5.0	13.56
angle, 90 degrees	nr	12.57	2.5	12.88
angle, 135 degrees	nr	21.86	2.5	22.41
outlet for 63mm pipe	nr	10.23	2.5	10.49
outlet for 76mm pipe	nr	20.03	2.5	20.53
outlet for 72 x 72mm pipe	nr	18.64	2.5	19.11
outlet for 102 x 76mm pipe	nr	18.64	2.5	19.11
stop end	nr	4.81	2.5	4.93
fascia bracket	nr	4.81	2.5	4.93
Beaded deep run, 113 x 75mm, polyester powder-coated gutter	m	18.16	5.0	19.07
angle, 90 degrees	nr	14.97	2.5	15.34
angle, 135 degrees	nr	24.25	2.5	24.86
outlet for 63mm pipe	nr	14.22	2.5	14.58
outlet for 76mm pipe	nr	23.51	2.5	24.10
outlet for 72 x 72mm pipe	nr	21.87	2.5	22.42
outlet for 102 x 76mm pipe	nr	21.87	2.5	22.42
stop end	nr	5.96	2.5	6.11
fascia bracket	nr	5.96	2.5	6.11
Barclay box gutter, 100 x 135mm, plain gutter	m	17.64	5.0	18.52
angle, 90 degrees	nr	22.00	2.5	22.55
outlet for 63mm pipe	nr	22.18	2.5	22.73

	Unit	Material supply (£)	Material waste (%)	Total (£)
outlet for 72 x 72mm pipe	nr	22.18	2.5	22.73
stop end	nr	3.59	2.5	3.68
rafter bracket	nr	3.59	2.5	3.68
Barclay box gutter, 100 x 87mm, polyester powder-coated gutter	m	23.82	5.0	25.01
angle, 90 degrees	nr	23.31	2.5	23.89
outlet for 63mm pipe	nr	9.96	2.5	10.21
outlet for 76 x 72mm pipe	nr	26.31	2.5	26.97
stop end	nr	4.76	2.5	4.88
rafter bracket	nr	4.77	2.5	4.89
GX box gutter, 100 x 75mm, plain	m	10.75	5.0	11.29
outlet for 63mm pipe	nr	19.78	2.5	20.27
outlet for 76mm pipe	nr	19.78	2.5	20.27
outlet for 72 x 72mm pipe	nr	19.78	2.5	20.27
outlet for 102 x 76mm pipe	nr	19.78	2.5	20.27
angle, 90 degrees	nr	19.78	2.5	20.27
angle, 135 degrees	nr	19.78	2.5	20.27
stop end	nr	4.67	2.5	4.79
fascia bracket	nr	3.04	2.5	3.12
GX box gutter, 100 x 75mm, polyester powder-coated	m	14.15	5.0	14.86
outlet for 63mm pipe	nr	23.45	2.5	24.04
outlet for 76mm pipe	nr	23.45	2.5	24.04
outlet for 72 x 72mm pipe	nr	23.45	2.5	24.04
outlet for 102 x 76mm pipe	nr	23.45	2.5	24.04
angle, 90 degrees	nr	23.45	2.5	24.04
angle, 135 degrees	nr	23.45	2.5	24.04
stop end	nr	7.09	2.5	7.27
fascia bracket	nr	4.07	2.5	4.17
GX box gutter, 125 x 100mm, plain	m	13.97	5.0	14.67
outlet for 63mm pipe	nr	24.67	2.5	25.29
outlet for 76mm pipe	nr	24.67	2.5	25.29
outlet for 102mm pipe	nr	24.67	2.5	25.29
outlet for 72 x 72mm pipe	nr	24.67	2.5	25.29
outlet for 102 x 76mm pipe	nr	24.67	2.5	25.29
outlet for 102 x 102mm pipe	nr	24.67	2.5	25.29

Gutters (cont'd)	Unit	Material supply (£)	Material waste (%)	Total (£)
angle, 90 degrees	nr	24.67	2.5	25.29
angle, 135 degrees	nr	24.67	2.5	25.29
stop end	nr	5.68	2.5	5.82
fascia bracket	nr	3.64	2.5	3.73
GX box gutter, 125 x 100mm, polyester powder-coated	m	17.92	5.0	18.82
outlet for 63mm pipe	nr	29.61	2.5	30.35
outlet for 76mm pipe	nr	29.61	2.5	30.35
outlet for 72 x 72mm pipe	nr	29.61	2.5	30.35
outlet for 102 x 76mm pipe	nr	29.61	2.5	30.35
outlet for 102 x 102mm pipe	nr	29.61	2.5	30.35
angle, 90 degrees	nr	29.61	2.5	30.35
angle, 135 degrees	nr	29.61	2.5	30.35
stop end	nr	8.13	2.5	8.33
fascia bracket	nr	4.84	2.5	4.96
GX box gutter, 150 x 100mm, plain	m	15.05	5.0	15.80
outlet for 63mm pipe	nr	25.11	2.5	25.74
outlet for 76mm pipe	nr	25.11	2.5	25.74
outlet for 102mm pipe	nr	25.11	2.5	25.74
outlet for 72 x 72mm pipe	nr	25.11	2.5	25.74
outlet for 102 x 76mm pipe	nr	25.11	2.5	25.74
outlet for 102 x 102mm pipe	nr	25.11	2.5	25.74
angle, 90 degrees	nr	25.11	2.5	25.74
angle, 135 degrees	nr	25.11	2.5	25.74
stop end	nr	6.27	2.5	6.43
fascia bracket	nr	0.34	2.5	0.35
GX box gutter, 150 x 100mm, polyester powder-coated	m	19.71	5.0	20.70
outlet for 63mm pipe	nr	30.00	2.5	30.75
outlet for 76mm pipe	nr	30.00	2.5	30.75
outlet for 102mm pipe	nr	30.00	2.5	30.75
outlet for 72 x 72mm pipe	nr	30.00	2.5	30.75
outlet for 102 x 76mm pipe	nr	30.00	2.5	30.75
outlet for 102 x 102mm pipe	nr	30.00	2.5	30.75
angle, 90 degrees	nr	30.00	2.5	30.75
angle, 135 degrees	nr	30.00	2.5	30.75
stop end	nr	8.77	2.5	8.99
fascia bracket	nr	5.68	2.5	5.82

MATERIAL COSTS

	Unit	Material supply (£)	Material waste (%)	Total (£)
Cast iron				
100mm half-round gutter	m	6.52	5.0	6.85
running outlet	nr	4.52	2.5	4.63
stop end	nr	1.52	2.5	1.56
stop end outlet	nr	3.22	2.5	3.30
angle, 90 degrees	nr	4.52	2.5	4.63
fascia bracket	nr	1.15	2.5	1.18
115mm half-round gutter	m	6.79	5.0	7.13
running outlet	nr	4.99	2.5	5.11
stop end	nr	2.13	2.5	2.18
stop end outlet	nr	3.91	2.5	4.01
angle, 90 degrees	nr	4.99	2.5	5.11
fascia bracket	nr	1.19	2.5	1.22
125mm half-round gutter	m	7.96	5.0	8.36
running outlet	nr	5.89	2.5	6.04
stop end	nr	2.13	2.5	2.18
stop end outlet	nr	4.51	2.5	4.62
angle, 90 degrees	nr	5.89	2.5	6.04
fascia bracket	nr	1.19	2.5	1.22
150mm half-round	m	13.60	5.0	14.28
running outlet	nr	8.10	2.5	8.30
stop end	nr	2.84	2.5	2.91
stop end outlet	nr	4.99	2.5	5.11
angle, 90 degrees	nr	8.10	2.5	8.30
fascia bracket	nr	1.34	2.5	1.37
100mm ogee	m	7.13	5.0	7.49
running outlet	nr	4.02	2.5	4.12
stop end	nr	1.36	2.5	1.39
stop end outlet	nr	3.03	2.5	3.11
angle, 90 degrees	nr	4.02	2.5	4.12
angle, 45 degrees	nr	4.02	2.5	4.12
fascia bracket	nr	1.48	2.5	1.52

Gutters (cont'd)	Unit	Material supply (£)	Material waste (%)	Total (£)
115mm ogee gutter	m	7.99	5.0	8.39
running outlet	nr	4.61	2.5	4.73
stop end	nr	1.96	2.5	2.01
stop end outlet	nr	3.22	2.5	3.30
angle, 90 degrees	nr	4.61	2.5	4.73
angle, 45 degrees	nr	4.61	2.5	4.73
fascia bracket	nr	1.48	2.5	1.52
125mm ogee gutter	m	8.38	5.0	8.80
running outlet	nr	5.62	2.5	5.76
stop end	nr	1.96	2.5	2.01
stop end outlet	nr	4.02	2.5	4.12
angle, 90 degrees	nr	5.48	2.5	5.62
angle, 45 degrees	nr	5.48	2.5	5.62
fascia bracket	nr	1.66	2.5	1.70

	Unit	Material supply (£)	Material waste (%)	Total (£)
ANCILLARIES				
Roof and balcony outlets				
uPVC Terrain				
flat roof outlet, small diameter				
55mm	nr	13.29	2.5	13.62
82mm	nr	13.29	2.5	13.62
flat roof outlet, large diameter				
82mm	nr	26.03	2.5	26.68
110mm	nr	26.03	2.5	26.68
domed roof outlet and small diameter				
82mm	nr	13.29	2.5	13.62
110mm	nr	13.29	2.5	13.62
domed roof outlet, large diameter				
55mm	nr	26.03	2.5	26.68
82mm	nr	26.03	2.5	26.68
combined roof outlet and vent pipe				
82mm	nr	37.63	2.5	38.57
110mm	nr	37.63	2.5	38.57
inverted roof outlet				
82mm	nr	57.30	2.5	58.73
110mm	nr	57.30	2.5	58.73
balcony outlet for screed finish				
82mm	nr	22.13	2.5	22.68
balcony outlet for asphalt finish				
82mm	nr	22.13	2.5	22.68
adaptor	nr	1.65	2.5	1.69
pressure release vent	nr	12.21	2.5	12.52

Ancillaries (cont'd)	Unit	Material supply (£)	Material waste (%)	Total (£)
uPVC Marley outlets				
flat roof outlet				
68mm	nr	18.57	2.5	19.03
160mm	nr	63.78	2.5	65.37
balcony outlet				
68mm	nr	18.57	2.5	19.03
flat roof grating				
82mm	nr	6.69	2.5	6.86
110mm	nr	6.69	2.5	6.86
balcony grating				
82mm	nr	6.69	2.5	6.86
110mm	nr	6.69	2.5	6.86
straight connectors				
82mm	nr	3.06	2.5	3.14
110mm	nr	3.26	2.5	3.34
bent connectors				
82mm	nr	7.00	2.5	7.18
110mm	nr	7.28	2.5	7.46
uPVC rainwater heads				
Terrain				
249 x 131 x 203mm	nr	8.30	2.5	8.51
280 x 121 x 194mm	nr	8.20	2.5	8.41
254 x 180 x 221mm	nr	26.10	2.5	26.75
254 x 180 x 222mm	nr	16.40	2.5	16.81
Marley				
280 x 155 x 230mm	nr	10.29	2.5	10.55
406 x 375 x 248mm	nr	53.22	2.5	54.55
Osma				
254 x 178 x 145mm	nr	6.56	2.5	6.72
Hepworth				
267 x 140 x 114mm	nr	7.18	2.5	7.36

	Unit	Material supply (£)	Material waste (%)	Total (£)
Cast aluminium (Fulbora Marketing)				
51mm, Mark 3, flat grating				
vertical spigot	nr	36.40	2.5	37.31
standard spigot	nr	37.20	2.5	38.13
vertical screw	nr	37.20	2.5	38.13
45 degree spigot	nr	47.75	2.5	48.94
45 degree screw	nr	59.20	2.5	60.68
90 degree spigot	nr	47.60	2.5	48.79
90 degree screw	nr	59.55	2.5	61.04
balcony	nr	51.25	2.5	52.53
76mm, Mark 3, flat grating				
vertical spigot	nr	48.55	2.5	49.76
standard spigot	nr	48.65	2.5	49.87
vertical screw	nr	48.55	2.5	49.76
45 degree spigot	nr	62.40	2.5	63.96
45 degree screw	nr	62.65	2.5	64.22
90 degree spigot	nr	65.25	2.5	66.88
90 degree screw	nr	65.50	2.5	67.14
two-way	nr	65.00	2.5	66.63
balcony	nr	53.55	2.5	54.89
101mm, Mark 3, flat grating				
vertical spigot	nr	68.70	2.5	70.42
standard spigot	nr	74.05	2.5	75.90
vertical screw	nr	73.75	2.5	75.59
45 degree spigot	nr	77.95	2.5	79.90
45 degree screw	nr	88.40	2.5	90.61
90 degree spigot	nr	92.35	2.5	94.66
90 degree screw	nr	90.80	2.5	93.07
two-way	nr	68.40	2.5	70.11
balcony	nr	63.50	2.5	65.09
152mm, Mark 3, flat grating				
vertical spigot	nr	85.85	2.5	88.00
standard spigot	nr	90.95	2.5	93.22
vertical screw	nr	87.05	2.5	89.23
90 degree screw	nr	99.95	2.5	102.45
two-way	nr	94.70	2.5	97.07
200mm, Mark 3, flat grating				
standard spigot	nr	89.45	2.5	91.69

Ancillaries (cont'd)	Unit	Material supply (£)	Material waste (%)	Total (£)
51mm, Mark 3, domed grating				
vertical spigot	nr	36.40	2.5	37.31
standard spigot	nr	37.20	2.5	38.13
vertical screw	nr	37.20	2.5	38.13
45 degree spigot	nr	47.75	2.5	48.94
45 degree screw	nr	59.20	2.5	60.68
90 degree spigot	nr	47.60	2.5	48.79
90 degree screw	nr	59.55	2.5	61.04
two-way	nr	49.95	2.5	51.20
gulley	nr	49.95	2.5	51.20
76mm, Mark 3, domed grating				
vertical spigot	nr	47.90	2.5	49.10
standard spigot	nr	48.20	2.5	49.41
vertical screw	nr	47.90	2.5	49.10
45 degree spigot	nr	61.70	2.5	63.24
45 degree screw	nr	62.10	2.5	63.65
90 degree spigot	nr	64.70	2.5	66.32
90 degree screw	nr	65.00	2.5	66.63
two-way	nr	64.70	2.5	66.32
101mm, Mark 3, domed grating				
vertical spigot	nr	62.95	2.5	64.52
standard spigot	nr	68.35	2.5	70.06
vertical screw	nr	68.90	2.5	70.62
45 degree spigot	nr	73.05	2.5	74.88
45 degree screw	nr	83.60	2.5	85.69
90 degree spigot	nr	87.60	2.5	89.79
90 degree screw	nr	89.25	2.5	91.48
two-way	nr	68.10	2.5	69.80
152mm, Mark 3, domed grating				
vertical spigot	nr	80.95	2.5	82.97
standard spigot	nr	85.45	2.5	87.59
vertical screw	nr	82.65	2.5	84.72
90 degree screw	nr	98.05	2.5	100.50
two-way	nr	94.10	2.5	96.45
200mm, Mark 3, domed grating				
standard spigot	nr	85.20	2.5	87.33

MATERIAL COSTS

	Unit	Material supply (£)	Material waste (%)	Total (£)
Roof outlets, polyurethene, Essergully, connections to pipes (Fulbora Marketing Ltd)				
100mm straight, insulated	nr	59.02	2.5	60.50
100mm straight, heated	nr	75.12	2.5	77.00
100mm angled, insulated	nr	77.81	2.5	79.76
100mm angled, heated	nr	91.23	2.5	93.51
PVC screw-threaded connectors				
51mm	nr	9.00	2.5	9.22
76mm	nr	15.00	2.5	15.38
101mm	nr	19.50	2.5	19.99
152mm	nr	23.00	2.5	23.57
PVC heat-shrink connectors				
51mm	nr	5.40	2.5	5.54
76mm	nr	5.90	2.5	6.05
101mm	nr	6.45	2.5	6.61

	Unit	Labour hours	Net labour (£)	Net material (£)	O'heads /profit (£)	Total (£)
PIPES						
uPVC Terrain 2100 system straight, connector joints to brickwork with pipe and fitting clips at 2m maximum centres, plugging						
69mm diameter pipe	m	0.25	2.40	5.29	1.15	8.84
Extra over for						
bend, 92.5 degrees	nr	0.15	1.44	2.31	0.56	4.31
bend, 112.5 degrees	nr	0.15	1.44	2.06	0.53	4.03
branch	nr	0.20	1.92	4.89	1.02	7.83
shoe	nr	0.30	2.88	1.39	0.64	4.91
access pipe	nr	0.25	2.40	9.78	1.83	14.01
uPVC Terrain 2200 square system, straight connector joints to brickwork with pipe fitting clips at 2m maximum centres, plugging						
62mm square pipe	m	0.25	2.40	6.80	1.38	10.58
Extra over for						
bend, 92.5 degrees	nr	0.15	1.44	2.28	0.56	4.28
branch	nr	0.20	1.92	5.54	1.12	8.58
shoe	nr	0.30	2.88	1.68	0.68	5.24
offset	nr	0.20	1.92	3.45	0.81	6.18
access pipe	nr	0.25	2.40	13.49	2.38	18.27
change piece, square to round adaptor	nr	0.10	0.96	2.62	0.54	4.12
connection to back inlet gulley, cement mortar (1:3)	nr	0.15	1.44	0.31	0.26	2.01
connection to drain pipe, adaptor	nr	0.15	1.44	6.00	1.12	8.56

	Unit	Labour hours	Net labour (£)	Net material (£)	O'heads /profit (£)	Total (£)
uPVC Terrain 2200 square system, straight connector joints to brickwork with pipe fitting clips at 2m maximum centres, plugging						
75mm square pipe	m	0.27	2.59	9.04	1.74	13.37
Extra over for						
bend, 112.5 degrees	nr	0.16	1.54	4.17	0.86	6.57
shoe	nr	0.32	3.07	2.15	0.78	6.00
change piece, 2300 to 2200 square pipe	nr	0.12	1.15	3.14	0.64	4.93
offset	nr	0.22	2.11	9.26	1.71	13.08
access pipe	nr	0.27	2.59	16.04	2.79	21.42
connection to back inlet gulley cement mortar (1:3)	nr	0.17	1.63	0.31	0.29	2.23
connection to drain pipe, adaptor square to round	nr	0.17	1.63	6.00	1.14	8.77
uPVC Terrain 82 system rainwater pipes straight half-round, solvent welded coupling joints, to brickwork with uPVC holderbats at 2m maximum centres, plugging						
82mm diameter pipe	m	0.33	3.17	8.01	1.68	12.86
Extra over for						
shoe	nr	0.35	3.36	6.63	1.50	11.49
bend, 92.5 degrees	nr	0.30	2.88	6.89	1.47	11.24
bend, 135 degrees	nr	0.30	2.88	6.89	1.47	11.24
branch, 92.5 degrees	nr	0.35	3.36	9.57	1.94	14.87
branch, 135 degrees	nr	0.35	3.36	9.57	1.94	14.87
spigot socket bend, 92.5 degrees	nr	0.30	2.88	6.63	1.43	10.94
seal ring adaptor	nr	0.20	1.92	1.74	0.55	4.21
connection to drain pipe, caulking bush	nr	0.25	2.40	5.84	1.24	9.48

Pipes (cont'd)	Unit	Labour hours	Net labour (£)	Net material (£)	O'heads /profit (£)	Total (£)
uPVC, Marley, Deepflow system, gutter union joints, to timber with support brackets at 1m maximum centres						
110mm semi-elliptical	m	0.30	2.88	3.98	1.03	7.89
Extra over for						
angle, 45 degrees	nr	0.28	2.69	3.54	0.93	7.16
angle, 90 degrees	nr	0.28	2.69	5.31	1.20	9.20
outlet	nr	0.18	1.73	3.24	0.75	5.72
stop end	nr	0.18	1.73	1.73	0.52	3.98
stop end outlet	nr	0.18	1.73	2.84	0.69	5.26
uPVC, Marley, Premier system, gutter union joints, to timber with support brackets at 1m maximum centres						
112mm half-round gutter	m	0.26	2.50	3.57	0.91	6.98
Extra over for						
angle, 45 degrees	nr	0.26	2.50	3.81	0.95	7.26
angle, 90 degrees	nr	0.26	2.50	3.63	0.92	7.05
outlet	nr	0.26	2.50	3.32	0.87	6.69
stop end	nr	0.14	1.34	3.32	0.70	5.36
stop end, outlet	nr	0.16	1.54	3.32	0.73	5.59
adaptor to ogee	nr	0.28	2.69	9.41	1.81	13.91
uPVC, Marley, Flowline system, gutter union joints, to timber with support brackets at 1m maximum centres						
116 x 60mm rectangular gutter	m	0.26	2.50	2.86	0.80	6.16
Extra over for						
angle, 45 degrees	nr	0.26	2.50	4.70	1.08	8.28
angle, 90 degrees	nr	0.26	2.50	3.22	0.86	6.58
outlet	nr	0.26	2.50	3.22	0.86	6.58
stop end, external	nr	0.16	1.54	1.42	0.44	3.40

	Unit	Labour hours	Net labour (£)	Net material (£)	O'heads /profit (£)	Total (£)
68mm diameter	m	0.25	2.40	4.31	1.01	7.72
Extra over for						
bend, 45 degrees	nr	0.15	1.44	3.01	0.67	5.12
bend, 67.5 degrees	nr	0.15	1.44	3.01	0.67	5.12
bend, 87.5 degrees	nr	0.15	1.44	3.01	0.67	5.12
offset bend, 67.5 degrees	nr	0.15	1.44	1.67	0.47	3.58
offset bend, 20 degrees	nr	0.15	1.44	1.81	0.49	3.74
branch, 67.5 degrees	nr	0.15	1.44	6.01	1.12	8.57
access pipe	nr	0.25	2.40	8.25	1.60	12.25
reducer, 110 to 68mm	nr	0.15	1.44	2.59	0.60	4.63
reducer, 82 to 68mm	nr	0.15	1.44	2.12	0.53	4.09
shoe	nr	0.30	2.88	2.59	0.82	6.29
65mm square pipe	m	0.25	2.40	4.98	1.11	8.49
Extra over for						
bend, 67.5 degrees	nr	0.15	1.44	1.95	0.51	3.90
bend, 87.5 degrees	nr	0.15	1.44	2.23	0.55	4.22
offset 54mm	nr	0.15	1.44	3.40	0.73	5.57
branch, 67.5 degrees	nr	0.15	1.44	7.49	1.34	10.27
access pipe	nr	0.25	2.40	10.38	1.92	14.70
shoe	nr	0.30	2.88	2.22	0.77	5.87
drain adaptor	nr	0.15	1.44	1.66	0.46	3.56
82mm diameter pipe	m	0.28	2.69	9.42	1.82	13.93
Extra over for						
bend, 87.5 degrees	nr	0.18	1.73	6.58	1.25	9.56
bend, adjustable	nr	0.18	1.73	11.04	1.92	14.69
offset bend, socket 67.5 degrees	nr	0.18	1.73	6.58	1.25	9.56
offset bend, spigot 67.5 degrees	nr	0.18	1.73	5.57	1.09	8.39
branch, adjustable	nr	0.18	1.73	10.32	1.81	13.86
access pipe	nr	0.28	2.69	12.66	2.30	17.65
shoe	nr	0.32	3.07	7.39	1.57	12.03
adaptor	nr	0.15	1.44	4.26	0.85	6.55
110mm diameter pipe	m	0.30	2.88	9.58	1.87	14.33
Extra over for						
bend, 87.5 degrees	nr	0.20	1.92	8.70	1.59	12.21
bend, adjustable	nr	0.20	0.00	11.04	1.66	12.70

Pipes (cont'd)	Unit	Labour hours	Net labour (£)	Net material (£)	O'heads /profit (£)	Total (£)
offset bend, end socket 67.5 degrees	nr	0.20	1.92	7.32	1.39	10.63
branch, 87.5 degrees	nr	0.20	1.92	12.23	2.12	16.27
access pipe, straight	nr	0.25	2.40	13.84	2.44	18.68
shoe	nr	0.34	3.26	9.55	1.92	14.73
adaptor	nr	0.15	1.44	3.58	0.75	5.77
160mm diameter pipe	m	0.35	3.36	21.59	3.74	28.69
Extra over for						
bend, adjustable	nr	0.24	2.30	21.89	3.63	27.82
offset bend, socket 67.5 degrees	nr	0.24	2.30	21.39	3.55	27.24
offset bend, spigot 67.5 degrees	nr	0.24	2.30	19.61	3.29	25.20
branch, 87.5 degrees	nr	0.24	2.30	34.25	5.48	42.03
access pipe	nr	0.28	2.69	27.89	4.59	35.17
shoe	nr	0.40	3.84	20.36	3.63	27.83
adaptor	nr	0.15	1.44	11.41	1.93	14.78
uPVC, Osma Minifit system, straight, connector fixing bracket joints, to brickwork with brackets at 2m maximum centres, plugging						
55mm square pipe	m	0.20	1.92	2.34	0.64	4.90
Extra over for						
bend, 45 degrees	nr	0.14	1.34	1.69	0.45	3.48
bend, 87.5 degrees	nr	0.14	1.34	1.96	0.49	3.79
offset bend	nr	0.14	1.34	1.26	0.39	2.99
shoe	nr	0.18	1.73	1.26	0.45	3.44
uPVC, Osma Roundline system, straight, connector joints, to brickwork with pipe brackets at 2m centres, plugging						
68mm diameter pipe	m	0.25	2.40	2.84	0.79	6.03
Extra over for						
bend, 87.5 degrees	nr	0.15	1.44	2.72	0.62	4.78
offset bend	nr	0.15	1.44	1.52	0.44	3.40
shoe	nr	0.30	2.88	1.60	0.67	5.15

	Unit	Labour hours	Net labour (£)	Net material (£)	O'heads /profit (£)	Total (£)
branch, 67.5 degrees	nr	0.20	1.92	4.82	1.01	7.75
access pipe	nr	0.15	1.44	6.62	1.21	9.27
universal connector	nr	0.20	1.92	2.51	0.66	5.09
ıPVC, Osma Squareline system						
51mm square pipe	m	0.25	2.40	2.91	0.80	6.11
Extra over for						
bend, 87.5 degrees	nr	0.15	1.44	2.76	0.63	4.83
offset bend	nr	0.15	1.44	1.59	0.45	3.48
shoe	nr	0.30	2.88	2.01	0.73	5.62
branch, 67.5 degrees	nr	0.20	1.92	4.88	1.02	7.82
access pipe	nr	0.15	1.44	6.71	1.22	9.37
connection to back inlet gulley cement mortar	nr	0.15	1.44	0.31	0.26	2.01
connection to drain, universal connector	nr	0.20	1.92	2.51	0.66	5.09
ıPVC, Osma Amazon system						
70mm square pipe	m	0.25	2.40	3.27	0.85	6.52
Extra over for						
offset bend	nr	0.15	1.44	1.66	0.46	3.56
shoe	nr	0.30	2.88	2.11	0.75	5.74
branch, 67.5 degrees	nr	0.20	1.92	5.31	1.08	8.31
access pipe	nr	0.15	1.44	7.52	1.34	10.30
offset	nr	0.15	1.44	2.50	0.59	4.53
offset with access	nr	0.15	1.44	7.52	1.34	10.30
connection to back inlet gulley cement mortar (1:3)	nr	0.15	1.44	0.31	0.26	2.01
connection to drain, universal connector joint	nr	0.20	1.92	2.51	0.66	5.09
ıPVC, Hepworth system, straight, onnector joints, to brickwork t 2m maximum centres, plugging						
0mm diameter pipe	m	0.22	2.11	3.47	0.84	6.42
Extra over for						
offset bend, 112.5 degrees	nr	0.15	1.44	1.97	0.51	3.92
shoe	nr	0.30	2.88	1.93	0.72	5.53

Pipes (cont'd)	Unit	Labour hours	Net labour (£)	Net material (£)	O'heads /profit (£)	Total (£)
68mm diameter pipe	m	0.25	2.40	3.42	0.87	6.69
Extra over for						
offset bend, 92.5 degrees	nr	0.15	1.44	2.89	0.65	4.98
offset bend, 112.5 degrees	nr	0.15	1.44	1.47	0.44	3.35
branch, 112.5 degrees	nr	0.15	1.44	5.25	1.00	7.69
shoe	nr	0.30	2.88	1.47	0.65	5.00
access pipe	nr	0.15	1.44	6.89	1.25	9.58
65mm square pipe	m	0.20	1.92	4.12	0.91	6.95
Extra over for						
shoe	nr	0.18	1.73	2.23	0.59	4.55
offset bend, 112.5 degrees	nr	0.14	1.34	1.66	0.45	3.45
mini offset, 44mm	nr	0.14	1.34	4.03	0.81	6.18
adaptor, square to round	nr	0.14	1.34	2.47	0.57	4.38
branch, 112.5 degrees	nr	0.14	1.34	5.34	1.00	7.68
access pipe	nr	0.14	1.34	6.57	1.19	9.10
connection to back inlet gulley cement mortar (1:3)	nr	0.14	1.34	0.31	0.25	1.90
Aluminium, Alumasc, straight, plain eared, spigot and socket dry joints to brickwork, plugging						
63.5mm diameter pipe	m	0.28	2.69	12.09	2.22	17.00
Extra over for						
offset, 76mm	nr	0.23	2.21	10.97	1.98	15.16
offset, 114mm	nr	0.23	2.21	12.08	2.14	16.43
offset, 152mm	nr	0.23	2.21	13.60	2.37	18.18
offset, 229mm	nr	0.23	2.21	15.43	2.65	20.29
offset, 305mm	nr	0.23	2.21	17.26	2.92	22.39
offset, 381mm	nr	0.23	2.21	19.14	3.20	24.55
offset, 457mm	nr	0.23	2.21	20.96	3.48	26.65
offset, 533mm	nr	0.23	2.21	22.82	3.75	28.78
offset, 610mm	nr	0.23	2.21	24.68	4.03	30.92
offset, 686mm	nr	0.23	2.21	26.48	4.30	32.99
offset, 762mm	nr	0.23	2.21	28.26	4.57	35.04
two-part offset, 228mm	nr	0.25	2.40	12.35	2.21	16.96
two-part offset, 381mm	nr	0.25	2.40	13.30	2.35	18.05
two-part offset, 457mm	nr	0.25	2.40	13.79	2.43	18.62
two-part offset, 685mm	nr	0.25	2.40	15.35	2.66	20.41
two-part offset, 914mm	nr	0.25	2.40	16.64	2.86	21.90

	Unit	Labour hours	Net labour (£)	Net material (£)	O'heads /profit (£)	Total (£)
shoe	nr	0.33	3.17	4.97	1.22	9.36
branch, 92.5 degrees	nr	0.23	2.21	7.47	1.45	11.13
branch, 112.5 degrees	nr	0.23	2.21	5.70	1.19	9.10
bend, 92.5 degrees	nr	0.18	1.73	5.70	1.11	8.54
bend, 112.5 degrees	nr	0.18	1.73	5.70	1.11	8.54
bend, 135 degrees	nr	0.18	1.73	5.70	1.11	8.54
76.5mm diameter pipe	m	0.33	3.17	10.75	2.09	16.01
Extra over for						
offset, 76mm	nr	0.28	2.69	13.29	2.40	18.38
offset, 114mm	nr	0.28	2.69	13.88	2.49	19.06
offset, 152mm	nr	0.28	2.69	15.28	2.70	20.67
offset, 229mm	nr	0.28	2.69	17.06	2.96	22.71
offset, 305mm	nr	0.28	2.69	18.93	3.24	24.86
offset, 381mm	nr	0.28	2.69	20.78	3.52	26.99
offset, 457mm	nr	0.28	2.69	24.29	4.05	31.03
offset, 533mm	nr	0.28	2.69	26.25	4.34	33.28
offset, 610mm	nr	0.28	2.69	28.27	4.64	35.60
offset, 686mm	nr	0.28	2.69	30.32	4.95	37.96
offset, 762mm	nr	0.28	2.69	32.22	5.24	40.15
two-part offset, 228mm	nr	0.30	2.88	13.35	2.43	18.66
two-part offset, 381mm	nr	0.30	2.88	14.49	2.61	19.98
two-part offset, 457mm	nr	0.30	2.88	15.12	2.70	20.70
two-part offset, 685mm	nr	0.30	2.88	17.06	2.99	22.93
two-part offset, 914mm	nr	0.30	2.88	18.72	3.24	24.84
shoe	nr	0.38	3.65	6.91	1.58	12.14
branch, 92.5 degrees	nr	0.28	2.69	9.29	1.80	13.78
branch, 112.5 degrees	nr	0.28	2.69	9.31	1.80	13.80
bend, 92.5 degrees	nr	0.28	2.69	7.47	1.52	11.68
bend, 112.5 degrees	nr	0.23	2.21	7.47	1.45	11.13
bend, 135 degrees	nr	0.23	2.21	16.99	2.88	22.08
102mm diameter pipe	m	0.36	3.46	18.75	3.33	25.54
Extra over for						
offset, 76mm	nr	0.31	2.98	15.04	2.70	20.72
offset, 114mm	nr	0.31	2.98	16.20	2.88	22.06
offset, 152mm	nr	0.31	2.98	17.47	3.07	23.52
offset, 229mm	nr	0.31	2.98	19.76	3.41	26.15
offset, 305mm	nr	0.31	2.98	22.08	3.76	28.82
offset, 381mm	nr	0.31	2.98	24.36	4.10	31.44
offset, 457mm	nr	0.31	2.98	27.89	4.63	35.50
offset, 533mm	nr	0.31	2.98	30.77	5.06	38.81

Pipes (cont'd)	Unit	Labour hours	Net labour (£)	Net material (£)	O'heads /profit (£)	Total (£)
offset, 610mm	nr	0.31	2.98	32.73	5.36	41.07
offset, 686mm	nr	0.31	2.98	35.12	5.71	43.81
offset, 762mm	nr	0.31	2.98	37.53	6.08	46.59
shoe	nr	0.41	3.94	8.25	1.83	14.02
branch, 92.5 degrees	nr	0.31	2.98	12.94	2.39	18.31
branch, 112.5 degrees	nr	0.31	2.98	12.94	2.39	18.31
bend, 92.5 degrees	nr	0.26	2.50	10.92	2.01	15.43
bend, 112.5 degrees	nr	0.26	2.50	10.92	2.01	15.43
bend, 135 degrees	nr	0.26	2.50	20.69	3.48	26.67
72mm square pipe	m	0.28	2.69	25.23	4.19	32.11
Extra over for						
offset, 76mm	nr	0.23	2.21	26.00	4.23	32.44
offset, 305mm	nr	0.23	2.21	34.94	5.57	42.72
offset, 533mm	nr	0.23	2.21	37.53	5.96	45.70
offset, 762mm	nr	0.23	2.21	40.05	6.34	48.60
shoe	nr	0.33	3.17	19.43	3.39	25.99
bend, 112.5 degrees	nr	0.23	2.21	20.93	3.47	26.61
rodding eye	nr	0.18	1.73	29.73	4.72	36.18
102 x 76mm rectangular pipe	m	0.30	2.88	23.25	3.92	30.05
Extra over for						
offset, 76mm	nr	0.25	2.40	29.87	4.84	37.11
offset, 305mm	nr	0.25	2.40	39.54	6.29	48.23
offset, 533mm	nr	0.25	2.40	42.84	6.79	52.03
offset, 762mm	nr	0.25	2.40	45.97	7.26	55.63
shoe	nr	0.35	3.36	20.93	3.64	27.93
bend, 112.5 degrees	nr	0.23	2.21	23.92	3.92	30.05
rodding eye	nr	0.20	1.92	31.40	5.00	38.32
102mm square pipe	m	0.33	3.17	28.06	4.68	35.91
Extra over for						
offset, 76mm	nr	0.28	2.69	34.16	5.53	42.38
offset, 305mm	nr	0.28	2.69	35.81	5.78	44.28
offset, 533mm	nr	0.28	2.69	51.34	8.10	62.13
offset, 762mm	nr	0.28	2.69	56.08	8.82	67.59
shoe	nr	0.38	3.65	23.46	4.07	31.18
bend, 112.5 degrees	nr	0.23	2.21	28.18	4.56	34.95
rodding eye	nr	0.23	2.21	35.41	5.64	43.26

	Unit	Labour hours	Net labour (£)	Net material (£)	O'heads /profit (£)	Total (£)
Cast iron straight, ears cast on, spigot and socket dry joints, primed, to brickwork, plugging						
50mm diameter pipe	m	0.25	2.40	15.91	2.75	21.06
Extra over for						
bend	nr	0.15	1.44	7.13	1.29	9.86
offset, 75mm	nr	0.20	1.92	10.89	1.92	14.73
offset, 230mm	nr	0.20	1.92	10.89	1.92	14.73
shoe	nr	0.30	2.88	11.63	2.18	16.69
connection to back inlet gulley cement mortar (1:3)	nr	0.15	1.44	0.34	0.27	2.05
65mm diameter pipe	m	0.28	2.69	16.54	2.88	22.11
Extra over for						
bend	nr	0.18	1.73	7.13	1.33	10.19
offset, 75mm	nr	0.23	2.21	10.89	1.96	15.07
offset, 230mm	nr	0.23	2.21	10.89	1.96	15.07
shoe	nr	0.33	3.17	11.63	2.22	17.02
connection to back inlet gulley cement mortar (1:3)	nr	0.15	1.44	0.34	0.27	2.05
75mm diameter pipe	m	0.35	3.36	16.54	2.98	22.88
Extra over for						
bend	nr	0.25	2.40	8.25	1.60	12.25
offset, 75mm	nr	0.30	2.88	10.89	2.07	15.84
offset, 230mm	nr	0.30	2.88	10.89	2.07	15.84
shoe	nr	0.40	3.84	11.63	2.32	17.79
connection to back inlet gulley cement mortar (1:3)	nr	0.17	1.63	0.34	0.30	2.27
100mm diameter pipe	m	0.40	3.84	21.37	3.78	28.99
Extra over for						
bend	nr	0.30	2.88	11.64	2.18	16.70
offset, 75mm	nr	0.35	3.36	18.64	3.30	25.30
offset, 230mm	nr	0.35	3.36	10.89	2.14	16.39
shoe	nr	0.45	4.32	15.12	2.92	22.36
connection to back inlet gulley, cement mortar (1:3)	nr	0.20	1.92	0.34	0.34	2.60

	Unit	Labour hours	Net labour (£)	Net material (£)	O'heads /profit (£)	Total (£)
GUTTERS						
uPVC, Terrain 2100 system, straight half-round, joint bracket joints, to timber with support brackets at 1m maximum centres						
110mm gutter	m	0.26	2.50	3.60	0.91	7.02
Extra over for						
running outlet	nr	0.26	2.50	2.94	0.82	6.26
angle, 90 degrees	nr	0.26	2.50	2.97	0.82	6.29
angle, 120 degrees	nr	0.26	2.50	3.52	0.90	6.92
angle, 135 degrees	nr	0.26	2.50	3.06	0.83	6.39
short stop end	nr	0.14	1.34	1.45	0.42	3.21
long stop end	nr	0.14	1.34	2.84	0.63	4.81
uPVC adaptor to cast iron half-round	nr	0.28	2.69	4.29	1.05	8.03
aluminium adaptor to cast iron half-round	nr	0.28	2.69	16.82	2.93	22.44
uPVC adaptor to cast iron ogee	nr	0.28	2.69	4.35	1.06	8.10
uPVC, Terrain 2200 square system, straight, joint bracket joints, to timber with support brackets at 1m maximum centres						
120mm gutter	m	0.26	2.50	3.77	0.94	7.21
Extra over for						
running outlet	nr	0.26	2.50	3.18	0.85	6.53
angle, 90 degrees	nr	0.26	2.50	2.97	0.82	6.29
angle, 135 degrees	nr	0.26	2.50	3.26	0.86	6.62
short stop end	nr	0.14	1.34	1.59	0.44	3.37
long stop end	nr	0.14	1.34	3.42	0.71	5.47
stop end outlet	nr	0.16	1.54	3.42	0.74	5.70
aluminium adaptor to cast iron ogee	nr	0.28	2.69	15.90	2.79	21.38

	Unit	Labour hours	Net labour (£)	Net material (£)	O'heads /profit (£)	Total (£)
uPVC, Terrain 2300 large-capacity system, straight, joint bracket joints, to timber with support brackets at 1m maximum centres						
132mm gutter	m	0.30	2.88	8.03	1.64	12.55
Extra over for						
running outlet	nr	0.28	2.69	8.78	1.72	13.19
angle, 90 degrees	nr	0.28	2.69	8.16	1.63	12.48
angle, 135 degrees	nr	0.28	2.69	15.98	2.80	21.47
stop end	nr	0.16	1.54	3.03	0.69	5.26
screwed outlet adaptor	nr	0.40	3.84	4.62	1.27	9.73
uPVC, Marley, Miniline system, gutter union joints, to timber with support brackets at 1m maximum centres						
75mm half-round gutter	m	0.26	2.50	1.82	0.65	4.97
Extra over for						
angle, 45 degrees	nr	0.26	2.50	2.31	0.72	5.53
angle, 90 degrees	nr	0.26	2.50	2.31	0.72	5.53
outlet	nr	0.26	2.50	1.94	0.67	5.11
stop end, internal	nr	0.14	1.34	0.84	0.33	2.51
stop end, external	nr	0.16	1.54	0.84	0.36	2.74
uPVC, Marley, Industrial system, gutter union joints, to timber with support brackets at 1m maximum centres						
150mm half-round gutter	m	0.30	2.88	8.02	1.64	12.54
Extra over for						
angle, 90 degrees	nr	0.28	2.69	10.84	2.03	15.56
outlet	nr	0.28	2.69	9.17	1.78	13.64
stop end, internal	nr	0.16	1.54	2.10	0.55	4.19
stop end, external	nr	0.16	1.54	2.95	0.67	5.16

Gutters (cont'd)	Unit	Labour hours	Net labour (£)	Net material (£)	O'heads /profit (£)	Total (£)
uPVC, Marley, Deepflow system, gutter union joints, to timber with support brackets at 1m maximum centres						
110mm semi-elliptical	m	0.30	2.88	3.98	1.03	7.89
Extra over for						
angle, 45 degrees	nr	0.28	2.69	3.54	0.93	7.16
angle, 90 degrees	nr	0.28	2.69	5.31	1.20	9.20
outlet	nr	0.18	1.73	3.24	0.75	5.72
stop end	nr	0.18	1.73	1.73	0.52	3.98
stop end outlet	nr	0.18	1.73	2.84	0.69	5.26
uPVC, Marley, Premier system, gutter union joints, to timber with support brackets at 1m maximum centres						
112mm half-round gutter	m	0.26	2.50	3.57	0.91	6.98
Extra over for						
angle, 45 degrees	nr	0.26	2.50	3.81	0.95	7.26
angle, 90 degrees	nr	0.26	2.50	3.63	0.92	7.05
outlet	nr	0.26	2.50	3.32	0.87	6.69
stop end	nr	0.14	1.34	3.32	0.70	5.36
stop end, outlet	nr	0.16	1.54	3.32	0.73	5.59
adaptor to ogee	nr	0.28	2.69	9.41	1.81	13.91
uPVC, Marley, Flowline system, gutter union joints, to timber with support brackets at 1m maximum centres						
116 x 60mm rectangular gutter	m	0.26	2.50	2.86	0.80	6.16
Extra over for						
angle, 45 degrees	nr	0.26	2.50	4.70	1.08	8.28
angle, 90 degrees	nr	0.26	2.50	3.22	0.86	6.58
outlet	nr	0.26	2.50	3.22	0.86	6.58
stop end, external	nr	0.16	1.54	1.42	0.44	3.40

	Unit	Labour hours	Net labour (£)	Net material (£)	O'heads /profit (£)	Total (£)
uPVC, Marley, Classic system, gutter union joints, to timber with support brackets at 1m maximum centres						
120 x 75mm ogee gutter	m	0.30	2.88	5.04	1.19	9.11
Extra over for						
angle, internal	nr	0.28	2.69	7.98	1.60	12.27
angle, external	nr	0.28	2.69	7.63	1.55	11.87
outlet	nr	0.28	2.69	6.25	1.34	10.28
stop end	nr	0.16	1.54	1.64	0.48	3.66
uPVC, Osma Minifit system, joint brackets joints, to timber with support brackets at 1m maximum centres						
75mm half-round gutter	m	0.24	2.30	1.91	0.63	4.84
Extra over for						
running outlet	nr	0.24	2.30	1.65	0.59	4.54
angle, 45 degrees	nr	0.24	2.30	1.81	0.62	4.73
angle, 90 degrees	nr	0.24	2.30	1.61	0.59	4.50
stop end	nr	0.14	1.34	0.72	0.31	2.37
stop end outlet	nr	0.14	1.34	1.61	0.44	3.39
uPVC, Osma Roundline system, joint bracket joints, to timber with support brackets at 1m maximum centres						
112mm half-round gutter	m	0.26	2.50	2.95	0.82	6.27
Extra over for						
swivelock running outlet	nr	0.26	2.50	2.65	0.77	5.92
swivelock running outlet with offset bend	nr	0.26	2.50	3.70	0.93	7.13
angle, 45 degrees	nr	0.26	2.50	2.75	0.79	6.04
angle, 60 degrees	nr	0.26	2.50	2.75	0.79	6.04
angle, 90 degrees	nr	0.26	2.50	2.49	0.75	5.74
stop end, internal	nr	0.14	1.34	0.67	0.30	2.31
stop end, external	nr	0.14	1.34	1.07	0.36	2.77

Gutters (cont'd)	Unit	Labour hours	Net labour (£)	Net material (£)	O'heads /profit (£)	Total (£)
connector to half-round cast iron gutter	nr	0.28	2.69	2.19	0.73	5.61
connector to ogee cast iron gutter	nr	0.28	2.69	3.47	0.92	7.08
uPVC, Osma Superline system, joint bracket joints, to timber with support brackets at 1m maximum centres						
125mm half-round gutter	m	0.26	2.50	3.47	0.90	6.87
Extra over for						
running outlet	nr	0.26	2.50	2.75	0.79	6.04
stop end, internal	nr	0.14	1.34	0.88	0.33	2.55
stop end, external	nr	0.14	1.34	1.34	0.40	3.08
angle, 45 degrees	nr	0.26	2.50	3.21	0.86	6.57
angle, 90 degrees	nr	0.26	2.50	3.05	0.83	6.38
uPVC, Osma Roofline system, joint bracket joints, to timber with support brackets at 1m maximum centres						
150mm half-round gutter	m	0.30	2.88	7.34	1.53	11.75
Extra over for						
running outlet	nr	0.30	2.88	6.89	1.47	11.24
stop end, internal	nr	0.16	1.54	1.09	0.39	3.02
stop end, external	nr	0.16	1.54	3.42	0.74	5.70
angle, 90 degrees	nr	0.30	2.88	8.94	1.77	13.59
uPVC, Osma Amazon system, joint bracket joints, to timber with support brackets at 1m maximum centres						
110 x 75mm rectangular gutter	m	0.30	2.88	3.48	0.95	7.31
Extra over for						
running outlet	nr	0.30	2.88	2.84	0.86	6.58
stop end, internal	nr	0.16	1.54	1.09	0.39	3.02
stop end, external	nr	0.16	1.54	1.53	0.46	3.53

	Unit	Labour hours	Net labour (£)	Net material (£)	O'heads /profit (£)	Total (£)
angle, 45 degrees	nr	0.30	2.88	2.99	0.88	6.75
angle, 90 degrees	nr	0.30	2.88	2.85	0.86	6.59
uPVC, Osma Squareline system, joint bracket joints, to timber with support brackets at 1m maximum centres						
100mm gutter	m	0.26	2.50	2.71	0.78	5.99
Extra over for						
running outlet	nr	0.26	2.50	2.72	0.78	6.00
stop end, internal	nr	0.14	1.34	0.85	0.33	2.52
stop end, external	nr	0.14	1.34	1.21	0.38	2.93
angle, 45 degrees	nr	0.26	2.50	2.77	0.79	6.06
angle, 90 degrees	nr	0.26	2.50	2.63	0.77	5.90
uPVC, Hepworth system, straight, joint bracket joints, to timber with support brackets at 1m maximum centres						
76mm half-round gutter	m	0.26	2.50	2.16	0.70	5.36
Extra over for						
stop end outlet	nr	0.14	1.34	2.27	0.54	4.15
running outlet	nr	0.26	2.50	2.27	0.72	5.49
stop end	nr	0.14	1.34	0.97	0.35	2.66
angle, 90 degrees	nr	0.26	2.50	2.29	0.72	5.51
112mm half-round gutter	m	0.26	2.50	2.82	0.80	6.12
Extra over for						
running outlet	nr	0.26	2.50	1.99	0.67	5.16
stop end	nr	0.14	1.34	1.11	0.37	2.82
angle, 90 degrees	nr	0.26	2.50	2.29	0.72	5.51
angle, 120 degrees	nr	0.26	2.50	2.62	0.77	5.89
angle, 135 degrees	nr	0.26	2.50	2.62	0.77	5.89
adaptor to metal ogee	nr	0.26	2.50	4.64	1.07	8.21
150mm half-round gutter	m	0.30	2.88	6.88	1.46	11.22
Extra over for						
stop end outlet	nr	0.16	1.54	6.91	1.27	9.72

Gutters (cont'd)	Unit	Labour hours	Net labour (£)	Net material (£)	O'heads /profit (£)	Total (£)
running outlet	nr	0.30	2.88	7.19	1.51	11.58
stop end	nr	0.16	1.54	3.29	0.72	5.55
angle, 90 degrees	nr	0.30	2.88	9.48	1.85	14.21
112mm square system gutter	m	0.26	2.50	2.91	0.81	6.22
Extra over for						
running outlet	nr	0.26	2.50	2.90	0.81	6.21
stop end	nr	0.14	1.34	1.02	0.35	2.71
angle, 90 degrees	nr	0.26	2.50	2.81	0.80	6.11
angle, 120 degrees	nr	0.26	2.50	3.56	0.91	6.97
angle, 135 degrees	nr	0.26	2.50	3.56	0.91	6.97
115 x 75mm high-capacity gutter	m	0.30	2.88	3.29	0.93	7.10
Extra over for						
running outlet	nr	0.30	2.88	3.36	0.94	7.18
stop end	nr	0.16	1.54	1.60	0.47	3.61
angle, 90 degrees	nr	0.30	2.88	3.26	0.92	7.06
angle, 135 degrees	nr	0.30	2.88	4.85	1.16	8.89
adaptor to deep section gutter	nr	0.30	2.88	2.69	0.84	6.41
adaptor to square down pipe	nr	0.30	2.88	2.38	0.79	6.05
Aluminium, Alumasc, mastic joints, to timber with support brackets at 915mm maximum centres						
100mm plain half-round gutter	m	0.29	2.78	6.30	1.36	10.44
Extra over for						
angle, 90 degrees	nr	0.29	2.78	3.76	0.98	7.52
angle, 135 degrees	nr	0.29	2.78	3.76	0.98	7.52
outlet	nr	0.29	2.78	4.34	1.07	8.19
stop end outlet	nr	0.17	1.63	3.58	0.78	5.99
stop end	nr	0.17	1.63	1.82	0.52	3.97
100mm, plain half-round, polyester powder-coated gutter	m	0.29	2.78	10.73	2.03	15.54
Extra over for						
angle, 90 degrees	nr	0.29	2.78	5.79	1.29	9.86
angle, 135 degrees	nr	0.29	2.78	5.79	1.29	9.86

	Unit	Labour hours	Net labour (£)	Net material (£)	O'heads /profit (£)	Total (£)
outlet	nr	0. 29	2. 78	6. 28	1. 36	10. 42
stop end outlet	nr	0. 17	1. 63	5. 59	1. 08	8. 30
stop end	nr	0. 17	1. 63	2. 83	0. 67	5. 13
113mm plain, half-round gutter	m	0. 30	2. 88	6. 68	1. 43	10. 99
Extra over for						
angle, 90 degrees	nr	0. 30	2. 88	4. 31	1. 08	8. 27
angle, 135 degrees	nr	0. 30	2. 88	4. 31	1. 08	8. 27
outlet	nr	0. 30	2. 88	4. 63	1. 13	8. 64
stop end outlet	nr	0. 18	1. 73	4. 01	0. 86	6. 60
stop end	nr	0. 18	1. 73	1. 87	0. 54	4. 14
113mm plain half-round, polyester powder-coated gutter	m	0. 30	2. 88	11. 26	2. 12	16. 26
Extra over for						
angle, 90 degrees	nr	0. 30	2. 88	6. 54	1. 41	10. 83
angle, 135 degrees	nr	0. 30	2. 88	6. 54	1. 41	10. 83
outlet	nr	0. 30	2. 88	6. 84	1. 46	11. 18
stop end outlet	nr	0. 18	1. 73	6. 41	1. 22	9. 36
stop end	nr	0. 18	1. 73	2. 96	0. 70	5. 39
125mm plain, half-round gutter	m	0. 31	2. 98	7. 88	1. 63	12. 49
Extra over for						
angle, 90 degrees	nr	0. 31	2. 98	4. 82	1. 17	8. 97
angle, 135 degrees	nr	0. 31	2. 98	4. 82	1. 17	8. 97
outlet	nr	0. 31	2. 98	4. 96	1. 19	9. 13
stop end	nr	0. 20	1. 92	2. 42	0. 65	4. 99
125mm plain, half-round, polyester powder-coated gutter	m	0. 31	2. 98	12. 33	2. 30	17. 61
Extra over for						
angle, 90 degrees	nr	0. 31	2. 98	7. 27	1. 54	11. 79
angle, 135 degrees	nr	0. 31	2. 98	7. 27	1. 54	11. 79
outlet	nr	0. 31	2. 98	7. 41	1. 56	11. 95
stop end	nr	0. 20	1. 92	3. 62	0. 83	6. 37

Gutters (cont'd)	Unit	Labour hours	Net labour (£)	Net material (£)	O'heads /profit (£)	Total (£)
113mm, beaded, plain, half-round gutter	m	0.29	2.78	8.12	1.63	12.53
Extra over for						
angle, 90 degrees	nr	0.29	2.78	6.01	1.32	10.11
outlet	nr	0.29	2.78	6.32	1.36	10.46
stop end	nr	0.17	1.63	3.34	0.75	5.72
113mm, beaded, half-round, polyester powder-coated gutter	m	0.29	2.78	13.69	2.47	18.94
Extra over for						
angle, 90 degrees	nr	0.29	2.78	8.39	1.68	12.85
outlet	nr	0.29	2.78	8.71	1.72	13.21
stop end	nr	0.17	1.63	4.49	0.92	7.04
125mm, beaded, plain, half-round gutter	m	0.34	3.26	9.59	1.93	14.78
Extra over for						
angle, 90 degrees	nr	0.34	3.26	6.63	1.48	11.37
outlet	nr	0.34	3.26	6.80	1.51	11.57
stop end	nr	0.20	1.92	3.91	0.87	6.70
125mm, beaded, half-round, polyester powder-coated gutter	m	0.34	3.26	15.02	2.74	21.02
Extra over for						
angle, 90 degrees	nr	0.34	3.26	9.29	1.88	14.43
outlet	nr	0.34	3.26	9.39	1.90	14.55
stop end	nr	0.30	2.88	5.16	1.21	9.25
100mm, Victorian ogee, plain gutter	m	0.31	2.98	8.29	1.69	12.96
Extra over for						
angle, 90 degrees	nr	0.31	2.98	4.18	1.07	8.23
angle, 135 degrees	nr	0.31	2.98	4.18	1.07	8.23
stop end outlet	nr	0.31	2.98	3.73	1.01	7.72
stop end	nr	0.17	1.63	1.99	0.54	4.16

	Unit	Labour hours	Net labour (£)	Net material (£)	O'heads /profit (£)	Total (£)
100mm, Victorian ogee, polyester powder-coated gutter	m	0.31	2.98	13.39	2.46	18.83
Extra over for						
angle, 90 degrees	nr	0.31	2.98	6.20	1.38	10.56
angle, 135 degrees	nr	0.31	2.98	6.20	1.38	10.56
stop end outlet	nr	0.31	2.98	5.71	1.30	9.99
stop end	nr	0.17	1.63	2.97	0.69	5.29
113mm, Victorian ogee, plain gutter	m	0.31	2.98	9.61	1.89	14.48
Extra over for						
angle, 90 degrees	nr	0.31	2.98	5.18	1.22	9.38
angle, 135 degrees	nr	0.31	2.98	5.18	1.22	9.38
stop end outlet	nr	0.31	2.98	4.17	1.07	8.22
stop end	nr	0.17	1.63	2.10	0.56	4.29
113mm, Victorian ogee, polyester powder-coated gutter	m	0.31	2.98	14.90	2.68	20.56
Extra over for						
angle, 90 degrees	nr	0.31	2.98	7.41	1.56	11.95
angle, 135 degrees	nr	0.31	2.98	7.41	1.56	11.95
stop end outlet	nr	0.31	2.98	6.40	1.41	10.79
stop end	nr	0.17	1.63	3.20	0.72	5.55
125mm, Victorian ogee, plain gutter	m	0.34	3.26	10.95	2.13	16.34
Extra over for						
angle, 90 degrees	nr	0.34	3.26	6.19	1.42	10.87
angle, 135 degrees	nr	0.34	3.26	6.19	1.42	10.87
stop end outlet	nr	0.34	3.26	4.82	1.21	9.29
stop end	nr	0.20	1.92	2.27	0.63	4.82
125mm, Victorian ogee, polyester powder-coated gutter	m	0.34	3.26	16.45	2.96	22.67
Extra over for						
angle, 90 degrees	nr	0.34	3.26	8.64	1.78	13.68
angle, 135 degrees	nr	0.34	3.26	8.64	1.78	13.68
stop end outlet	nr	0.34	3.26	7.27	1.58	12.11
stop end	nr	0.20	1.92	3.48	0.81	6.21

Gutters (cont'd)	Unit	Labour hours	Net labour (£)	Net material (£)	O'heads /profit (£)	Total (£)
Moulded No. 46, 100 x 75mm, plain gutter	m	0.34	3.26	12.42	2.35	18.03
Extra over for						
angle, 90 degrees	nr	0.34	3.26	4.74	1.20	9.20
angle, 135 degrees	nr	0.34	3.26	4.74	1.20	9.20
outlet for 63mm pipe	nr	0.34	3.26	4.89	1.22	9.37
outlet for 76mm pipe	nr	0.34	3.26	14.40	2.65	20.31
outlet for 102 x 76mm pipe	nr	0.34	3.26	24.10	4.10	31.46
Moulded No. 46, 100 x 75mm, polyester powder-coated gutter	m	0.34	3.26	17.91	3.18	24.35
Extra over for						
angle, 90 degrees	nr	0.34	3.26	7.00	1.54	11.80
angle, 135 degrees	nr	0.34	3.26	7.00	1.54	11.80
outlet for 63mm pipe	nr	0.34	3.26	7.09	1.55	11.90
outlet for 76mm pipe	nr	0.34	3.26	16.63	2.98	22.87
outlet for 102 x 76mm pipe	nr	0.34	3.26	27.03	4.54	34.83
Moulded No. 46, 125 x 100mm, plain gutter	m	0.36	3.46	15.99	2.92	22.37
Extra over for						
angle, 90 degrees	nr	0.36	3.46	7.61	1.66	12.73
angle, 135 degrees	nr	0.36	3.46	7.61	1.66	12.73
outlet for 63mm pipe	nr	0.36	3.46	7.61	1.66	12.73
outlet for 76mm pipe	nr	0.36	3.46	7.61	1.66	12.73
outlet for 102 x 76mm pipe	nr	0.36	3.46	7.61	1.66	12.73
outlet for 102 x 102mm pipe	nr	0.36	3.46	29.17	4.89	37.52
Moulded No. 46, 125 x 100mm, polyester powder-coated gutter	m	0.36	3.46	21.78	3.79	29.03
Extra over for						
angle, 90 degrees	nr	0.36	3.46	10.08	2.03	15.57
angle, 135 degrees	nr	0.36	3.46	10.08	2.03	15.57
outlet for 63mm pipe	nr	0.36	3.46	10.08	2.03	15.57
outlet for 76mm pipe	nr	0.36	3.46	10.08	2.03	15.57
outlet for 102 x 76mm pipe	nr	0.36	3.46	10.08	2.03	15.57
outlet for 102 x 102mm pipe	nr	0.36	3.46	32.10	5.33	40.89

	Unit	Labour hours	Net labour (£)	Net material (£)	O'heads /profit (£)	Total (£)
Moulded No. 46, 150 x 100mm, plain gutter	m	0.38	3.65	20.72	3.66	28.03
Extra over for						
angle, 90 degrees	nr	0.38	3.65	17.60	3.19	24.44
angle, 135 degrees	nr	0.38	3.65	17.60	3.19	24.44
outlet for 63mm pipe	nr	0.38	3.65	17.60	3.19	24.44
outlet for 76mm pipe	nr	0.38	3.65	17.60	3.19	24.44
outlet for 102 x 76mm pipe	nr	0.38	3.65	33.73	5.61	42.99
outlet for 102 x 102mm pipe	nr	0.38	3.65	33.73	5.61	42.99
Moulded No. 46, 150 x 100mm, polyester powder-coated gutter	m	0.38	3.65	27.11	4.61	35.37
Extra over for						
angle, 90 degrees	nr	0.38	3.65	20.27	3.59	27.51
angle, 135 degrees	nr	0.38	3.65	20.27	3.59	27.51
outlet for 63mm pipe	nr	0.38	3.65	20.27	3.59	27.51
outlet for 76mm pipe	nr	0.38	3.65	20.27	3.59	27.51
outlet for 102 x 76mm pipe	nr	0.38	3.65	36.40	6.01	46.06
outlet for 102 x 102mm pipe	nr	0.38	3.65	36.40	6.01	46.06
Beaded deep run, 113 x 75mm, plain gutter	m	0.38	3.65	13.56	2.58	19.79
Extra over for						
angle, 90 degrees	nr	0.38	3.65	12.88	2.48	19.01
angle, 135 degrees	nr	0.38	3.65	22.41	3.91	29.97
outlet for 63mm pipe	nr	0.38	3.65	11.00	2.20	16.85
outlet for 76mm pipe	nr	0.38	3.65	20.53	3.63	27.81
outlet for 72 x 72mm pipe	nr	0.38	3.65	19.11	3.41	26.17
outlet for 102 x 76mm pipe	nr	0.38	3.65	19.11	3.41	26.17
stop end	nr	0.38	3.65	4.93	1.29	9.87
Beaded deep run, 113 x 75mm, polyester powder-coated gutter	m	0.38	3.65	19.07	3.41	26.13
Extra over for						
angle, 90 degrees	nr	0.38	3.65	15.34	2.85	21.84
angle, 135 degrees	nr	0.38	3.65	24.86	4.28	32.79
outlet for 63mm pipe	nr	0.38	3.65	14.58	2.73	20.96
outlet for 76mm pipe	nr	0.38	3.65	24.10	4.16	31.91
outlet for 72 x 72mm pipe	nr	0.38	3.65	22.42	3.91	29.98

Gutters (cont'd)	Unit	Labour hours	Net labour (£)	Net material (£)	O'heads /profit (£)	Total (£)
outlet for 102 x 76mm pipe	nr	0.38	3.65	22.42	3.91	29.98
stop end	nr	0.38	3.65	6.11	1.46	11.22
Barclay box gutter, 100 x 87mm, plain gutter	m	0.36	3.46	18.52	3.30	25.28
Extra over for						
angle, 90 degrees	nr	0.36	3.46	22.73	3.93	30.12
outlet for 63mm pipe	nr	0.36	3.46	7.77	1.68	12.91
outlet for 76mm pipe	nr	0.36	3.46	22.73	3.93	30.12
stop end	nr	0.22	2.11	3.68	0.87	6.66
Barclay box gutter 100 x 87mm, polyester powder-coated gutter, angle 90 degrees	m	0.36	3.46	25.01	4.27	32.74
Extra over for						
angle, 90 degrees	nr	0.36	3.46	23.89	4.10	31.45
outlet for 63mm pipe	nr	0.36	3.46	10.21	2.05	15.72
outlet for 76mm pipe	nr	0.36	3.46	26.97	4.56	34.99
stop end	nr	0.22	2.11	4.88	1.05	8.04
GX box gutter, 100 x 75mm, plain	m	0.34	3.26	11.29	2.18	16.73
Extra over for						
outlet for 63mm pipe	nr	0.34	3.26	20.27	3.53	27.06
outlet for 76mm pipe	nr	0.34	3.26	20.27	3.53	27.06
outlet for 72 x 72mm pipe	nr	0.34	3.26	20.27	3.53	27.06
outlet for 102 x 76mm pipe	nr	0.34	3.26	20.27	3.53	27.06
angle, 90 degrees	nr	0.34	3.26	20.27	3.53	27.06
angle, 135 degrees	nr	0.34	3.26	20.27	3.53	27.06
stop end	nr	0.20	1.92	4.79	1.01	7.72
GX box gutter, 100 x 75mm, polyester powder-coated	m	0.34	3.26	14.86	2.72	20.84
Extra over for						
outlet for 63mm pipe	nr	0.34	3.26	24.04	4.09	31.39
outlet for 76mm pipe	nr	0.34	3.26	24.04	4.09	31.39
outlet for 72 x 72mm pipe	nr	0.34	3.26	24.04	4.09	31.39
outlet for 102 x 76mm pipe	nr	0.34	3.26	24.04	4.09	31.39
angle, 90 degrees	nr	0.34	3.26	24.04	4.09	31.39
angle, 135 degrees	nr	0.34	3.26	24.04	4.09	31.39
stop end	nr	0.20	1.92	7.27	1.38	10.57

	Unit	Labour hours	Net labour (£)	Net material (£)	O'heads /profit (£)	Total (£)
GX box gutter, 125 x 100mm, plain	m	0.30	2.88	14.67	2.63	20.18
Extra over for						
outlet for 63mm pipe	nr	0.36	3.46	25.29	4.31	33.06
outlet for 76mm pipe	nr	0.36	3.46	25.29	4.31	33.06
outlet for 102mm pipe	nr	0.36	3.46	25.29	4.31	33.06
outlet for 72 x 72mm pipe	nr	0.36	3.46	25.29	4.31	33.06
outlet for 102 x 76mm pipe	nr	0.36	3.46	25.29	4.31	33.06
outlet for 102 x 102mm pipe	nr	0.36	3.46	25.29	4.31	33.06
angle, 90 degrees	nr	0.36	3.46	25.29	4.31	33.06
angle, 135 degrees	nr	0.36	3.46	25.29	4.31	33.06
stop end	nr	0.22	2.11	5.82	1.19	9.12
GX box gutter, 125 x 100mm, polyester powder-coated	m	0.36	3.46	18.82	3.34	25.62
Extra over for						
outlet for 63mm pipe	nr	0.36	3.46	30.35	5.07	38.88
outlet for 76mm pipe	nr	0.36	3.46	30.35	5.07	38.88
outlet for 72 x 72mm pipe	nr	0.36	3.46	30.35	5.07	38.88
outlet for 102 x 76mm pipe	nr	0.36	3.46	30.35	5.07	38.88
outlet for 102 x 102mm pipe	nr	0.36	3.46	30.35	5.07	38.88
angle, 90 degrees	nr	0.36	3.46	30.35	5.07	38.88
angle, 135 degrees	nr	0.36	3.46	30.35	5.07	38.88
stop end	nr	0.22	2.11	8.33	1.57	12.01
GX box gutter, 150 x 100mm, plain	m	0.38	3.65	15.80	2.92	22.37
Extra over for						
outlet for 63mm pipe	nr	0.38	3.65	25.74	4.41	33.80
outlet for 76mm pipe	nr	0.38	3.65	25.74	4.41	33.80
outlet for 102mm pipe	nr	0.38	3.65	25.74	4.41	33.80
outlet for 72 x 72mm pipe	nr	0.38	3.65	25.74	4.41	33.80
outlet for 102 x 76mm pipe	nr	0.38	3.65	25.74	4.41	33.80
outlet for 102 x 102mm pipe	nr	0.38	3.65	25.74	4.41	33.80
angle, 90 degrees	nr	0.38	3.65	25.74	4.41	33.80
angle, 135 degrees	nr	0.38	3.65	25.74	4.41	33.80
stop end	nr	0.24	2.30	6.43	1.31	10.04
GX box gutter, 150 x 100mm, polyester powder-coated	m	0.38	3.65	20.70	3.65	28.00
Extra over for						
outlet for 63mm pipe	nr	0.38	3.65	30.75	5.16	39.56

Gutters (cont'd)	Unit	Labour hours	Net labour (£)	Net material (£)	O'heads /profit (£)	Total (£)
outlet for 76mm pipe	nr	0.38	3.65	30.75	5.16	39.56
outlet for 102mm pipe	nr	0.38	3.65	30.75	5.16	39.56
outlet for 72 x 72mm pipe	nr	0.38	3.65	30.75	5.16	39.56
outlet for 102 x 76mm pipe	nr	0.38	3.65	30.75	5.16	39.56
outlet for 102 x 102mm pipe	nr	0.38	3.65	30.75	5.16	39.56
angle, 90 degrees	nr	0.38	3.65	30.75	5.16	39.56
angle, 135 degrees	nr	0.38	3.65	30.75	5.16	39.56
stop end	nr	0.24	2.30	8.99	1.69	12.98
Cast iron, straight mastic joints, primed, to timber with brackets at 1m maximum centres						
100mm half-round gutter	m	0.35	3.36	6.71	1.51	11.58
Extra over for						
running outlet	nr	0.35	3.36	4.52	1.18	9.06
stop end	nr	0.20	1.92	1.52	0.52	3.96
stop end outlet	nr	0.20	1.92	3.22	0.77	5.91
angle, 90 degrees	nr	0.35	3.36	4.52	1.18	9.06
115mm half-round gutter	m	0.35	3.36	8.14	1.72	13.23
Extra over for						
running outlet	nr	0.35	3.36	4.99	1.25	9.60
stop end	nr	0.20	1.92	2.13	0.61	4.66
stop end outlet	nr	0.20	1.92	3.91	0.87	6.70
angle, 90 degrees	nr	0.35	3.36	4.99	1.25	9.60
125mm half-round gutter	m	0.40	3.84	9.53	2.01	15.38
Extra over for						
running outlet	nr	0.40	3.84	5.75	1.44	11.03
stop end	nr	0.25	2.40	2.13	0.68	5.21
stop end outlet	nr	0.25	2.40	4.51	1.04	7.95
angle, 90 degrees	nr	0.40	3.84	5.89	1.46	11.19
150mm half-round gutter	m	0.45	4.32	15.73	3.01	23.06
Extra over for						
running outlet	nr	0.45	4.32	8.10	1.86	14.28
stop end	nr	0.28	2.69	2.84	0.83	6.36
stop end outlet	nr	0.28	2.69	4.99	1.15	8.83
angle, 90 degrees	nr	0.45	4.32	8.10	1.86	14.28

	Unit	Labour hours	Net labour (£)	Net material (£)	O'heads /profit (£)	Total (£)
100mm ogee gutter	m	0.37	3.55	8.67	1.83	14.05
Extra over for						
running outlet	nr	0.37	3.55	4.02	1.14	8.71
stop end	nr	0.22	2.11	1.36	0.52	3.99
stop end outlet	nr	0.22	2.11	3.03	0.77	5.91
angle, 90 degrees	nr	0.37	3.55	4.02	1.14	8.71
angle, 45 degrees	nr	0.37	3.55	4.02	1.14	8.71
115mm ogee gutter	m	0.37	3.55	9.63	1.98	15.16
Extra over for						
running outlet	nr	0.37	3.55	4.61	1.22	9.38
stop end	nr	0.22	2.11	1.96	0.61	4.68
stop end outlet	nr	0.22	2.11	3.22	0.80	6.13
angle, 90 degrees	nr	0.37	3.55	4.61	1.22	9.38
angle, 45 degrees	nr	0.37	3.55	4.61	1.22	9.38
125mm ogee gutter	m	0.42	4.03	10.15	2.13	16.31
Extra over for						
running outlet	nr	0.42	4.03	5.48	1.43	10.94
stop end	nr	0.27	2.59	1.96	0.68	5.23
stop end outlet	nr	0.27	2.59	4.02	0.99	7.60
angle, 90 degrees	nr	0.42	4.03	5.48	1.43	10.94
angle, 45 degrees	nr	0.42	4.03	5.48	1.43	10.94

	Unit	Labour hours	Net labour (£)	Net material (£)	O'heads /profit (£)	Total (£)
ANCILLARIES						
Roof and balcony outlets						
uPVC Terrain						
flat roof outlet, small diameter						
55mm	nr	0.50	4.80	13.96	2.81	21.57
82mm	nr	0.60	5.76	13.96	2.96	22.68
flat roof outlet, large diameter						
82mm	nr	0.50	4.80	27.35	4.82	36.97
110mm	nr	0.60	5.76	27.35	4.97	38.08
domed roof outlet and small diameter						
82mm	nr	0.50	4.80	13.96	2.81	21.57
110mm	nr	0.60	5.76	13.96	2.96	22.68
domed roof outlet, large diameter						
55mm	nr	0.50	4.80	27.35	4.82	36.97
82mm	nr	0.60	5.76	27.35	4.97	38.08
combined roof outlet and vent pipe						
82mm	nr	0.60	5.76	39.53	6.79	52.08
110mm	nr	0.70	6.72	39.53	6.94	53.19
inverted roof outlet						
82mm	nr	0.60	5.76	60.20	9.89	75.85
110mm	nr	0.70	6.72	60.20	10.04	76.96
balcony outlet for screed finish						
82mm	nr	0.50	4.80	23.25	4.21	32.26
balcony outlet for asphalt finish						
82mm	nr	0.50	4.80	23.25	4.21	32.26
adaptor	nr	0.05	0.48	1.73	0.33	2.54
pressure release vent	nr	0.70	6.72	12.83	2.93	22.48

	Unit	Labour hours	Net labour (£)	Net material (£)	O'heads /profit (£)	Total (£)
uPVC Marley outlets						
flat roof outlet						
68mm	nr	0.50	4.80	19.51	3.65	27.96
160mm	nr	0.80	7.68	68.03	11.36	87.07
balcony outlet						
68mm	nr	0.50	4.80	19.51	3.65	27.96
flat roof grating						
82mm	nr	0.50	4.80	7.03	1.77	13.60
110mm	nr	0.80	7.68	7.03	2.21	16.92
balcony grating						
82mm	nr	0.50	4.80	7.03	1.77	13.60
110mm	nr	0.80	7.68	7.03	2.21	16.92
straight connectors						
82mm	nr	0.25	2.40	3.22	0.84	6.46
110mm	nr	0.30	2.88	3.22	0.91	7.02
bent connectors						
82mm	nr	0.30	2.88	7.36	1.54	11.78
110mm	nr	0.35	3.36	7.65	1.65	12.66
uPVC rainwater heads						
Terrain						
249 x 131 x 203mm	nr	0.40	3.84	8.72	1.88	14.44
280 x 121 x 194mm	nr	0.50	4.80	8.62	2.01	15.43
254 x 180 x 221mm	nr	0.50	4.80	27.42	4.83	37.05
254 x 180 x 222mm	nr	0.50	4.80	17.23	3.30	25.33
Marley						
280 x 155 x 230mm	nr	0.50	4.80	10.81	2.34	17.95
406 x 375 x 248mm	nr	0.70	6.72	55.91	9.39	72.02
Osma						
254 x 178 x 145mm	nr	0.40	3.84	6.89	1.61	12.34
Hepworth						
267 x 140 x 114mm	nr	0.50	4.80	7.54	1.85	14.19

Ancillaries (cont'd)	Unit	Labour hours	Net labour (£)	Net material (£)	O'heads /profit (£)	Total (£)
Cast aluminium (Fulbora Marketing)						
51mm, Mark 3, flat grating						
vertical spigot	nr	0.60	5.76	37.31	6.46	49.53
standard spigot	nr	0.60	5.76	38.13	6.58	50.47
vertical screw	nr	0.60	5.76	38.13	6.58	50.47
45 degree spigot	nr	0.60	5.76	48.94	8.21	62.91
45 degree screw	nr	0.60	5.76	60.68	9.97	76.41
90 degree spigot	nr	0.60	5.76	48.79	8.18	62.73
90 degree screw	nr	0.60	5.76	61.04	10.02	76.82
balcony	nr	0.60	5.76	52.53	8.74	67.03
76mm, Mark 3, flat grating						
vertical spigot	nr	0.70	6.72	49.76	8.47	64.95
standard spigot	nr	0.70	6.72	49.87	8.49	65.08
vertical screw	nr	0.70	6.72	49.76	8.47	64.95
45 degree spigot	nr	0.70	6.72	63.96	10.60	81.28
45 degree screw	nr	0.70	6.72	64.22	10.64	81.58
90 degree spigot	nr	0.70	6.72	66.88	11.04	84.64
90 degree screw	nr	0.70	6.72	67.14	11.08	84.94
two-way	nr	0.70	6.72	66.63	11.00	84.35
balcony	nr	0.70	6.72	54.89	9.24	70.85
101mm, Mark 3, flat grating						
vertical spigot	nr	0.80	7.68	70.42	11.71	89.81
standard spigot	nr	0.80	7.68	75.90	12.54	96.12
vertical screw	nr	0.80	7.68	75.59	12.49	95.76
45 degree spigot	nr	0.80	7.68	79.90	13.14	100.72
45 degree screw	nr	0.80	7.68	90.61	14.74	113.03
90 degree spigot	nr	0.80	7.68	94.66	15.35	117.69
90 degree screw	nr	0.80	7.68	93.07	15.11	115.86
two-way	nr	0.80	7.68	70.11	11.67	89.46
balcony	nr	0.80	7.68	65.09	10.92	83.69
152mm, Mark 3, flat grating						
vertical spigot	nr	1.00	9.60	88.00	14.64	112.24
standard spigot	nr	1.00	9.60	93.22	15.42	118.24
vertical screw	nr	1.00	9.60	89.23	14.82	113.65
90 degree screw	nr	1.00	9.60	102.45	16.81	128.86
two-way	nr	1.00	9.60	97.07	16.00	122.67
200mm, Mark 3, flat grating						
standard spigot	nr	1.20	11.52	91.69	15.48	118.69

	Unit	Labour hours	Net labour (£)	Net material (£)	O'heads /profit (£)	Total (£)
51mm, Mark 3, domed grating						
vertical spigot	nr	0.60	5.76	37.31	6.46	49.53
standard spigot	nr	0.60	5.76	38.13	6.58	50.47
vertical screw	nr	0.60	5.76	38.13	6.58	50.47
45 degree spigot	nr	0.60	5.76	48.94	8.21	62.91
45 degree screw	nr	0.60	5.76	60.68	9.97	76.41
90 degree spigot	nr	0.60	5.76	48.79	8.18	62.73
90 degree screw	nr	0.60	5.76	61.04	10.02	76.82
two-way	nr	0.60	5.76	51.20	8.54	65.50
gulley	nr	0.60	5.76	51.20	8.54	65.50
76mm, Mark 3, domed grating						
vertical spigot	nr	0.70	6.72	49.10	8.37	64.19
standard spigot	nr	0.70	6.72	49.41	8.42	64.55
vertical screw	nr	0.70	6.72	49.10	8.37	64.19
45 degree spigot	nr	0.70	6.72	63.24	10.49	80.45
45 degree screw	nr	0.70	6.72	63.65	10.56	80.93
90 degree spigot	nr	0.70	6.72	66.32	10.96	84.00
90 degree screw	nr	0.70	6.72	66.63	11.00	84.35
two-way	nr	0.70	6.72	66.32	10.96	84.00
101mm, Mark 3, domed grating						
vertical spigot	nr	0.80	7.68	64.52	10.83	83.03
standard spigot	nr	0.80	7.68	70.06	11.66	89.40
vertical screw	nr	0.80	7.68	70.62	11.75	90.05
45 degree spigot	nr	0.80	7.68	74.88	12.38	94.94
45 degree screw	nr	0.80	7.68	85.69	14.01	107.38
90 degree spigot	nr	0.80	7.68	89.79	14.62	112.09
90 degree screw	nr	0.80	7.68	91.48	14.87	114.03
two-way	nr	0.80	7.68	69.80	11.62	89.10
152mm, Mark 3, domed grating						
vertical spigot	nr	1.00	9.60	82.97	13.89	106.46
standard spigot	nr	1.00	9.60	87.59	14.58	111.77
vertical screw	nr	1.00	9.60	84.72	14.15	108.47
90 degree screw	nr	1.00	9.60	100.50	16.52	126.62
two-way	nr	1.00	9.60	96.45	15.91	121.96
200mm, Mark 3, domed grating						
standard spigot	nr	1.10	10.56	87.33	14.68	112.57

Ancillaries (cont'd)	Unit	Labour hours	Net labour (£)	Net material (£)	O'heads /profit (£)	Total (£)
Roof outlets, polyurethene, Essergully, connections to pipes (Fulbora Marketing Ltd)						
100mm straight, insulated	nr	1.00	9.60	60.50	10.51	80.61
100mm straight, heated	nr	1.00	9.60	77.00	12.99	99.59
100mm angled, insulated	nr	1.10	10.56	79.76	13.55	103.87
100mm angled, heated	nr	1.10	10.56	93.51	15.61	119.68
PVC screw-threaded connectors						
51mm	nr	0.25	2.40	9.22	1.74	13.36
76mm	nr	0.30	2.88	15.38	2.74	21.00
101mm	nr	0.30	2.88	19.99	3.43	26.30
152mm	nr	0.35	3.36	23.57	4.04	30.97
PVC heat-shrink connectors						
51mm	nr	0.25	2.40	5.54	1.19	9.13
76mm	nr	0.30	2.88	6.05	1.34	10.27
101mm	nr	0.30	2.88	6.61	1.42	10.91

	Unit	Material supply (£)	Material waste (%)	Total (£)

R11 FOUL DRAINS ABOVE GROUND

WASTE PIPES

MuPVC Terrain waste system 200, solvent-welded joints

	Unit	Material supply (£)	Material waste (%)	Total (£)
32mm diameter pipe	m	2.16	5.0	2.27
bend, 91.25 degrees	nr	1.73	2.5	1.77
bend, 135 degrees	nr	1.73	2.5	1.77
bend, 165 degrees	nr	1.83	2.5	1.88
tee, 91.25 degrees	nr	2.46	2.5	2.52
tee, 135 degrees	nr	2.64	2.5	2.71
connector, iron to MuPVC	nr	1.47	2.5	1.51
connector, copper to MuPVC	nr	1.95	2.5	2.00
caulking bush	nr	2.06	2.5	2.11
screw to wall bracket	nr	1.24	2.5	1.27
40mm diameter pipe	m	2.65	5.0	2.78
bend, 91.25 degrees	nr	1.95	2.5	2.00
bend, 135 degrees	nr	1.95	2.5	2.00
bend, 165 degrees	nr	2.16	2.5	2.21
tee, 91.25 degrees	nr	3.06	2.5	3.14
tee, 135 degrees	nr	3.20	2.5	3.28
sweep cross	nr	8.81	2.5	9.03
connector, iron to MuPVC	nr	1.74	2.5	1.78
connector, copper to MuPVC	nr	2.06	2.5	2.11
caulking bush	nr	2.06	2.5	2.11
screw to wall bracket	nr	1.47	2.5	1.51
50mm diameter pipe	m	3.89	5.0	4.08
bend, 91.25 degrees	nr	2.81	2.5	2.88
bend, 135 degrees	nr	2.89	2.5	2.96
bend, 165 degrees	nr	3.05	2.5	3.13
tee, 91.25 degrees	nr	5.35	2.5	5.48
tee, 135 degrees	nr	5.58	2.5	5.72
sweep cross	nr	8.91	2.5	9.13
connector, iron to MuPVC	nr	2.49	2.5	2.55
connector, copper to MuPVC	nr	2.76	2.5	2.83
caulking bush	nr	2.06	2.5	2.11
screw to wall bracket	nr	1.88	2.5	1.93

	Unit	Material supply (£)	Material waste (%)	Total (£)
MuPVC, Marley waste system, solvent-welded joints				
32mm diameter pipe	m	1.85	5.0	1.94
branch, 88.5 degrees	nr	1.48	2.5	1.52
branch, 45 degrees	nr	1.48	2.5	1.52
reducer, 32 to 22mm	nr	1.25	2.5	1.28
adaptor, iron to MuPVC	nr	1.48	2.5	1.52
pipe clip	nr	0.24	2.5	0.25
40mm diameter pipe	m	2.24	5.0	2.35
branch, 88.5 degrees	nr	1.68	2.5	1.72
branch, 45 degrees	nr	1.68	2.5	1.72
reducer, 40 to 32mm	nr	1.03	2.5	1.06
cross tee	nr	8.08	2.5	8.28
adaptor, iron to MuPVC	nr	1.68	2.5	1.72
pipe clip	nr	0.28	2.5	0.29
50mm diameter pipe	m	3.18	5.0	3.34
branch, 88.5 degrees	nr	2.35	2.5	2.41
branch, 45 degrees	nr	2.35	2.5	2.41
cross tee	nr	9.38	2.5	9.61
reducer, 50 to 40mm	nr	1.74	2.5	1.78
reducer, 50 to 32mm	nr	1.91	2.5	1.96
adaptor, iron to MuPVC	nr	2.50	2.5	2.56
caulking bush	nr	2.70	2.5	2.77
pipe clip	nr	0.81	2.5	0.83
Ultra ABS, Marley waste system, solvent-welded joints				
32mm diameter pipe	m	1.17	5.0	1.23
bend, 45 degrees	nr	0.95	2.5	0.97
bend, 88.5 degrees	nr	1.10	2.5	1.13
branch, 88.5 degrees	nr	1.63	2.5	1.67
branch, 45 degrees	nr	1.63	2.5	1.67
adaptor, iron to MuPVC	nr	1.31	2.5	1.34
pipe clip	nr	0.22	0.0	0.22

	Unit	Material supply (£)	Material waste (%)	Total (£)
40mm diameter pipe	m	1.45	5.0	1.52
bend, 88.5 degrees	nr	1.31	2.5	1.34
bend, 45 degrees	nr	1.10	2.5	1.13
branch, 88.5 degrees	nr	1.88	2.5	1.93
branch, 45 degrees	nr	1.88	2.5	1.93
cross tee	nr	3.42	2.5	3.51
reducer, 40 to 32mm	nr	0.41	2.5	0.42
adaptor, iron to MuPVC	nr	1.40	2.5	1.43
pipe clip	nr	0.28	2.5	0.29
50mm diameter pipe	m	1.78	5.0	1.87
bend, 88.5 degrees	nr	2.09	2.5	2.14
bend, 45 degrees	nr	1.88	2.5	1.93
branch, 88.5 degrees	nr	2.86	2.5	2.93
branch, 45 degrees	nr	2.86	2.5	2.93
cross tee	nr	4.67	2.5	4.79
reducer, 50 to 40mm	nr	0.93	2.5	0.95
reducer, 50 to 32mm	nr	0.93	2.5	0.95
adaptor, iron to MuPVC	nr	1.88	2.5	1.93
caulking bush	nr	2.70	2.5	2.77
pipe clip	nr	0.81	0.0	0.81
MuPVC, Marley waste system, push-fit joints				
32mm diameter pipe	m	0.59	5.0	0.62
bend, 90 degrees	nr	0.72	2.5	0.74
bend, 45 degrees	nr	0.75	2.5	0.77
branch, 88.5 degrees	nr	0.94	2.5	0.96
tank connector	nr	1.10	2.5	1.13
multi-fit waste connector	nr	1.41	2.5	1.45
pipe clip	nr	0.21	2.5	0.22
40mm diameter pipe	m	0.74	5.0	0.78
bend, 90 degrees	nr	0.75	2.5	0.77
bend, 45 degrees	nr	0.81	2.5	0.83
branch, 88.5 degrees	nr	1.21	2.5	1.24

Waste pipes (cont'd)	Unit	Material supply (£)	Material waste (%)	Total (£)
reducer, 40 to 32mm	nr	0.81	2.5	0.83
tank connector	nr	1.17	2.5	1.20
multi-fit waste connector	nr	1.46	2.5	1.50
pipe clip	nr	0.27	2.5	0.28
ABS, OsmaWeld waste system, solvent-welded joints				
32mm diameter pipe	m	1.08	5.0	1.13
bend, 90 degrees	nr	0.90	2.5	0.92
bend, 87.5 degrees	nr	0.95	2.5	0.97
bend, 45 degrees	nr	0.84	2.5	0.86
bend, long tail 90 degrees	nr	1.00	2.5	1.03
tee	nr	1.35	2.5	1.38
access plug	nr	0.78	2.5	0.80
connector, straight tank	nr	1.69	2.5	1.73
connector, universal	nr	1.21	2.5	1.24
connector, copper to ABS	nr	1.69	2.5	1.73
connector, iron to ABS	nr	1.42	2.5	1.46
reducer, 32 to 19mm	nr	0.66	2.5	0.68
pipe bracket	nr	0.20	0.0	0.20
pipe clip	nr	0.20	2.5	0.21
40mm diameter pipe	m	1.33	5.0	1.40
bend, 90 degrees	nr	1.07	2.5	1.10
bend, 87.5 degrees	nr	1.14	2.5	1.17
bend, 45 degrees	nr	0.95	2.5	0.97
bend, long tail 90 degrees	nr	1.20	2.5	1.23
tee	nr	1.65	2.5	1.69
cross tee	nr	3.13	2.5	3.21
access plug	nr	0.81	2.5	0.83
connector, straight tank	nr	1.84	2.5	1.89
connector, universal	nr	1.31	2.5	1.34
connector, copper to ABS	nr	1.84	2.5	1.89
connector, iron to ABS	nr	1.50	2.5	1.54
reducer, 40 to 32mm	nr	0.75	2.5	0.77
pipe bracket	nr	0.21	2.5	0.22
pipe clip	nr	0.21	2.5	0.22

	Unit	Material supply (£)	Material waste (%)	Total (£)
50mm diameter pipe	m	1.66	5.0	1.74
bend, 87.5 degrees	nr	1.72	2.5	1.76
bend, 45 degrees	nr	1.49	2.5	1.53
bend, long tail 87.5 degrees	nr	2.14	2.5	2.19
tee	nr	2.38	2.5	2.44
cross tee	nr	4.38	2.5	4.49
access plug	nr	1.35	2.5	1.38
connector, iron to ABS	nr	1.87	2.5	1.92
reducer, 50 to 40mm	nr	0.98	2.5	1.00
reducer, 50 to 32mm	nr	0.98	2.5	1.00
pipe bracket	nr	0.39	2.5	0.40
Polypropylene, Osma ClearBore ring seal joints				
32mm diameter pipe	m	0.72	5.0	0.76
bend, 90 degrees	nr	1.17	2.5	1.20
bend, 87.5 degrees	nr	1.19	2.5	1.22
bend, 45 degrees	nr	1.08	2.5	1.11
bend, spigot, 45 degrees	nr	1.03	2.5	1.06
tee	nr	1.70	2.5	1.74
access plug	nr	0.78	2.5	0.80
connector, straight tank	nr	1.69	2.5	1.73
connector, universal	nr	1.21	2.5	1.24
connector, copper to ABS	nr	1.49	2.5	1.53
connector, iron to ABS	nr	1.42	2.5	1.46
reducer, 32 to 19mm	nr	0.66	2.5	0.68
pipe bracket	nr	0.20	2.5	0.21
pipe clip	nr	0.20	2.5	0.21
40mm diameter pipe	m	0.90	5.0	0.95
bend, 90 degrees	nr	1.27	2.5	1.30
bend, 87.5 degrees	nr	1.34	2.5	1.37
bend, 45 degrees	nr	1.14	2.5	1.17
bend, spigot, 45 degrees	nr	1.08	2.5	1.11
tee	nr	1.96	2.5	2.01
access plug	nr	0.81	2.5	0.83
connector, straight tank	nr	1.84	2.5	1.89
connector, universal	nr	1.91	2.5	1.96
connector, copper to ABS	nr	1.70	2.5	1.74
connector, iron to ABS	nr	1.50	2.5	1.54

Waste pipes (cont'd)	Unit	Material supply (£)	Material waste (%)	Total (£)
reducer, 40 to 32mm	nr	0.74	2.5	0.76
pipe bracket	nr	0.21	2.5	0.22
pipe clip	nr	0.21	2.5	0.22
50mm diameter pipe	m	1.31	5.0	1.38
bend, 87.5 degrees	nr	2.34	2.5	2.40
bend, 45 degrees	nr	2.08	2.5	2.13
tee	nr	2.86	2.5	2.93
access plug	nr	1.35	2.5	1.38
connector, iron to ABS	nr	1.07	2.5	1.10
reducer, 50 to 40mm	nr	1.10	2.5	1.13
reducer, 50 to 32mm	nr	1.10	2.5	1.13
pipe bracket	nr	0.39	2.5	0.40
Polypropylene, Hepworth waste system, push-fit joints				
32mm diameter pipe	m	0.60	5.0	0.63
reducer, 32 to 21.5mm	nr	0.42	2.5	0.43
tank connector, to iron	nr	0.56	2.5	0.57
tank connector, to copper	nr	0.56	2.5	0.57
bend, 135 degrees	nr	0.56	2.5	0.57
bend, 150 degrees	nr	0.56	2.5	0.57
bend, 90 degrees	nr	0.56	2.5	0.57
bend, 92.5 degrees	nr	0.56	2.5	0.57
tee, 92.5 degrees	nr	0.56	2.5	0.57
plug, access	nr	0.56	2.5	0.57
pipe clip	nr	0.28	2.5	0.29
40mm diameter pipe	m	0.71	5.0	0.75
reducer, 40 to 21.5mm	nr	0.42	2.5	0.43
reducer, 40 to 32mm	nr	0.56	2.5	0.57
tank connector, to iron	nr	0.56	2.5	0.57
tank connector, to copper	nr	0.56	2.5	0.57
bend, 135 degrees	nr	0.56	2.5	0.57
bend, 150 degrees	nr	0.56	2.5	0.57
bend, 90 degrees	nr	0.56	2.5	0.57
bend, 92.5 degrees	nr	0.56	2.5	0.57
tee, 92.5 degrees	nr	0.56	2.5	0.57
plug, access	nr	0.56	2.5	0.57
pipe clip	nr	0.28	0.0	0.28

	Unit	Material supply (£)	Material waste (%)	Total (£)
50mm diameter pipe	m	1.20	5.0	1.26
reducer, 50 to 40mm	nr	1.00	2.5	1.03
reducer, 50 to 32mm	nr	1.00	2.5	1.03
bend, 135 degrees	nr	1.00	2.5	1.03
bend, 150 degrees	nr	1.00	2.5	1.03
bend, 90 degrees	nr	1.00	2.5	1.03
bend, 92.5 degrees	nr	1.00	2.5	1.03
tee, 92.5 degrees	nr	1.00	2.5	1.03
plug, access	nr	1.00	2.5	1.03
pipe clip	nr	0.33	2.5	0.34
ABS, Hepworth waste systems, solvent-welded joints				
32mm diameter pipe	m	1.00	5.0	1.05
bend, 135 degrees	nr	0.62	2.5	0.64
bend, 90 degrees	nr	0.62	2.5	0.64
bend, 92.5 degrees	nr	0.62	2.5	0.64
tee, 92.5 degrees	nr	0.62	2.5	0.64
plug, access	nr	0.62	2.5	0.64
pipe clip	nr	0.33	2.5	0.34
40mm diameter pipe	m	1.23	5.0	1.29
reducer, 40 to 32mm	nr	0.62	2.5	0.64
bend, 135 degrees	nr	0.62	2.5	0.64
bend, 90 degrees	nr	0.62	2.5	0.64
bend, 92.5 degrees	nr	0.62	2.5	0.64
tee, 92.5 degrees	nr	0.62	2.5	0.64
tee, cross	nr	0.62	2.5	0.64
plug, access	nr	0.62	2.5	0.64
pipe clip	nr	0.33	2.5	0.34

	Unit	Material supply (£)	Material waste (%)	Total (£)
SOIL PIPES				
uPVC, Terrain soil system 100, solvent-welded joints				
82mm diameter pipe	m	5.52	5.0	5.80
bend, 92.5 degrees	nr	7.12	2.5	7.30
bend, 135 degrees	nr	7.12	2.5	7.30
branch, single, 92.5 degrees	nr	9.90	2.5	10.15
branch, single, 135 degrees	nr	9.90	2.5	10.15
vent cowl	nr	1.99	2.5	2.04
weathered apron	nr	1.99	2.5	2.04
screwed to wall holderbat	nr	2.20	2.5	2.26
110mm diameter pipe	m	6.64	5.0	6.97
bend, 92.5 degrees	nr	9.75	2.5	9.99
bend, 135 degrees	nr	9.75	2.5	9.99
bend, variable, spigot/spigot	nr	13.88	2.5	14.23
bend, variable, spigot/socket	nr	13.88	2.5	14.23
bend, WC connecting, 92.5 degrees	nr	10.38	2.5	10.64
bend, access, 92.5 degrees	nr	21.87	2.5	22.42
branch, single, 92.5 degrees	nr	12.75	2.5	13.07
branch, single, 104 degrees	nr	12.90	2.5	13.22
branch, single, 135 degrees	nr	12.90	2.5	13.22
branch, single, access	nr	24.98	2.5	25.60
branch, double, 92.5 degrees	nr	30.07	2.5	30.82
branch, double, 104 degrees	nr	30.07	2.5	30.82
branch, double, 135 degrees	nr	31.43	2.5	32.22
branch, corner, 92.5 degrees	nr	54.40	2.5	55.76
bossed pipe connector, solvent welded sockets, single	nr	3.10	2.5	3.18
bossed pipe connector, spigot boss and solvent-welded sockets	nr	3.22	2.5	3.30
connector, WC horizontal outlet	nr	8.80	2.5	9.02
connector, WC variable	nr	14.82	2.5	15.19
vent cowl	nr	2.01	2.5	2.06
weathered apron	nr	2.34	2.5	2.40
screw to wall holderbat	nr	2.28	2.5	2.34

	Unit	Material supply (£)	Material waste (%)	Total (£)
160mm diameter pipe	m	14.15	5.0	14.86
bend, 92.5 degrees	nr	21.01	2.5	21.54
bend, 135 degrees	nr	19.89	2.5	20.39
branch, single, 92.5 degrees	nr	67.14	2.5	68.82
branch, single, 135 degrees	nr	43.13	2.5	44.21
branch, double, 92.5 degrees	nr	46.96	2.5	48.13
branch, double, 104 degrees	nr	52.39	2.5	53.70
vent cowl	nr	5.26	2.5	5.39
weathered apron	nr	7.07	2.5	7.25
screw to wall holderbat	nr	5.50	2.5	5.64
uPVC, Marley soil system, spigot and socket seal joints				
82mm diameter pipe	m	4.78	5.0	5.02
bend, spigot/socket	nr	7.02	2.5	7.20
bend, solvent socket	nr	7.02	2.5	7.20
bend, adjustable	nr	11.77	2.5	12.06
offset bend, socket	nr	7.02	2.5	7.20
offset bend, spigot	nr	5.93	2.5	6.08
branch, socket/socket/solvent socket, plain	nr	10.29	2.5	10.55
branch, adjustable socket/ spigot, 3 boss upstand	nr	11.01	2.5	11.29
boss branch, 3 boss upstand	nr	5.22	2.5	5.35
boss branch, single 50mm, 3 boss upstand/solvent sockets	nr	5.59	2.5	5.73
rear access bend, adjustable	nr	18.49	2.5	18.95
rear access branch, adjustable	nr	19.44	2.5	19.93
vent terminal	nr	1.89	2.5	1.94
vent slate cap	nr	2.56	2.5	2.62
WC connector, straight	nr	3.06	2.5	3.14
two piece pipe clip	nr	3.12	2.5	3.20
110mm diameter pipe	m	5.04	5.0	5.29
bend, spigot/socket	nr	9.28	2.5	9.51
bend, solvent socket	nr	9.28	2.5	9.51
bend, adjustable	nr	11.77	2.5	12.06
offset bend, end socket	nr	7.80	2.5	8.00
branch, single socket/spigot, plain	nr	14.29	2.5	14.65

Soil pipes (cont'd)	Unit	Material supply (£)	Material waste (%)	Total (£)
branch, single socket/spigot, 5 boss upstand	nr	12.08	2.5	12.38
branch, socket/socket/solvent socket, 3 boss upstand	nr	14.29	2.5	14.65
branch, socket/socket/solvent socket, plain	nr	14.29	2.5	14.65
branch, double socket/spigot, corner	nr	49.73	2.5	50.97
branch, double socket/spigot, 2 boss upstand/2 solvent sockets	nr	27.58	2.5	28.27
multi-branch	nr	34.23	2.5	35.09
multi-branch, with socket	nr	4.22	2.5	4.33
multi-branch, with access cap	nr	7.56	2.5	7.75
boss branch, collar boss	nr	14.68	2.5	15.05
boss branch, single 40mm	nr	6.42	2.5	6.58
boss branch, single 32mm	nr	6.42	2.5	6.58
boss branch, single 50mm, all solvent connections	nr	3.00	2.5	3.08
boss branch, single 40mm, double solvent sockets	nr	2.79	2.5	2.86
boss branch, single 32mm, double solvent sockets	nr	2.79	2.5	2.86
boss branch, 4 boss upstand	nr	6.10	2.5	6.25
boss branch, single 50mm, 3 boss upstand/solvent sockets	nr	5.57	2.5	5.71
rear access bend, plain	nr	21.62	2.5	22.16
rear access branch, 2 boss upstand	nr	25.45	2.5	26.09
roof cowl	nr	10.85	2.5	11.12
vent terminal	nr	2.10	2.5	2.15
weathering collar	nr	2.10	2.5	2.15
WC connector, straight	nr	3.26	2.5	3.34
adaptor, to salt-glazed socket	nr	4.93	2.5	5.05
adaptor, to cast iron socket	nr	11.95	2.5	12.25
two-piece pipe clip	nr	3.20	2.5	3.28
160mm diameter pipe	m	13.30	5.0	13.97
bend, spigot/socket	nr	22.17	2.5	22.72

	Unit	Material supply (£)	Material waste (%)	Total (£)
bend, solvent socket	nr	22.17	2.5	22.72
bend, adjustable	nr	22.35	2.5	22.91
offset bend, socket	nr	22.81	2.5	23.38
offset bend, spigot	nr	20.91	2.5	21.43
branch, single socket/spigot, plain	nr	36.52	2.5	37.43
branch, triple solvent socket, 3 boss upstand	nr	26.65	2.5	27.32
branch, socket/socket/solvent socket, plain	nr	29.13	2.5	29.86
rear access bend, adjustable	nr	33.86	2.5	34.71
vent terminal	nr	5.53	2.5	5.67
vent slate cap	nr	9.74	2.5	9.98
adaptor, to cast iron socket	nr	32.32	2.5	33.13
two-piece pipe clip	nr	4.83	2.5	4.95
uPVC, Osmasoil system spigot and socket ring seal joints				
82mm diameter pipe	m	4.63	5.0	4.86
reducer, 82 to 55mm	nr	4.94	2.5	5.06
connector, uPVC to cast iron	nr	4.67	2.5	4.79
bend, 87.5, 67.5, 45 degrees	nr	5.95	2.5	6.10
branch, 87.5 degrees	nr	9.21	2.5	9.44
branch, 67.5 degrees	nr	9.21	2.5	9.44
weathering collar	nr	1.95	2.5	2.00
balloon grating	nr	1.70	2.5	1.74
pipe bracket	nr	1.22	2.5	1.25
110mm diameter pipe	m	5.20	5.0	5.46
reducer, 110 to 82mm	nr	5.30	2.5	5.43
reducer, 110 to 55mm	nr	6.10	2.5	6.25
connector, uPVC to cast iron	nr	6.13	2.5	6.28
connector, uPVC clay	nr	3.48	2.5	3.57
bend, 87.5, 67.5, 45 degrees	nr	7.98	2.5	8.18
bend, 30 or 15 degrees	nr	7.68	2.5	7.87
bend, adjustable, 30 degrees	nr	8.77	2.5	8.99
branch, 87.5 degrees	nr	11.26	2.5	11.54
branch, 5 boss, 87.5 degrees	nr	11.71	2.5	12.00
branch, 67.5 degrees	nr	12.11	2.5	12.41
branch, 45 degrees	nr	11.75	2.5	12.04

Soil pipes (cont'd)	Unit	Material supply (£)	Material waste (%)	Total (£)
branch, 76 degrees	nr	12.11	2.5	12.41
branch, double, 45 degrees	nr	26.04	2.5	26.69
branch, corner, 87.5 degrees	nr	44.37	2.5	45.48
connector, WC, 90 degrees	nr	7.60	2.5	7.79
weathering collar	nr	1.92	2.5	1.97
balloon grating	nr	1.56	2.5	1.60
vent cowl	nr	8.63	2.5	8.85
pipe bracket	nr	1.38	2.5	1.41
160mm diameter pipe	m	12.22	5.0	12.83
reducer, 160 to 110mm	nr	10.86	2.5	11.13
connector, uPVC to cast iron	nr	12.56	2.5	12.87
bend, 87.5, 67.5, 45 degrees	nr	18.80	2.5	19.27
bend, offset, bottom	nr	19.29	2.5	19.77
bend, offset, top	nr	21.24	2.5	21.77
branch, equal, 87.5 degrees	nr	33.94	2.5	34.79
branch, unequal, 87.5 degrees	nr	26.17	2.5	26.82
branch, unequal double, 87.5 degrees	nr	46.79	2.5	47.96
weathering collar	nr	6.43	2.5	6.59
balloon grating	nr	4.87	2.5	4.99
pipe bracket	nr	3.39	2.5	3.47
Cast iron to BS416, Drainage System Ltd 'Timesaver' system flexible joints				
50mm diameter pipe	m	12.44	5.0	13.06
bend, short radius	nr	9.35	2.5	9.58
bend, short radius, access	nr	23.04	2.5	23.62
branch, 87.5 degrees	nr	14.06	2.5	14.41
branch, 67.5 degrees	nr	14.07	2.5	14.42
branch, 45 degrees	nr	14.07	2.5	14.42
branch, access, 87.5 degrees	nr	27.77	2.5	28.46
branch, access, 67.5 degrees	nr	27.77	2.5	28.46
branch, access, 45 degrees	nr	27.77	2.5	28.46
branch, double, 87.5 degrees	nr	23.67	2.5	24.26
branch, double access, 87.5 degrees	nr	37.36	2.5	38.29
branch, parallel	nr	12.74	2.5	13.06

	Unit	Material supply (£)	Material waste (%)	Total (£)
change piece, to soil	nr	7.93	2.5	8.13
offset, 75mm projection	nr	9.26	2.5	9.49
offset, 115mm projection	nr	11.52	2.5	11.81
offset, 150mm projection	nr	14.92	2.5	15.29
offset, 225mm projection	nr	14.92	2.5	15.29
offset, 300mm projection	nr	16.99	2.5	17.41
roof vent connector for asphalt	nr	24.11	2.5	24.71
roof vent connector for felt	nr	58.20	2.5	59.66
roof outlet	nr	31.24	2.5	32.02
wall fixing bracket	nr	5.10	2.5	5.23
75mm diameter pipe	m	12.44	5.0	13.06
bend, short radius	nr	9.35	2.5	9.58
bend, short radius, access	nr	22.34	2.5	22.90
bend, long radius	nr	17.65	2.5	18.09
bend, long radius, access	nr	31.35	2.5	32.13
branch, 87.5 degrees	nr	14.07	2.5	14.42
branch, 67.5 degrees	nr	14.07	2.5	14.42
branch, 45 degrees	nr	14.07	2.5	14.42
branch, access, 87.5 degrees	nr	27.77	2.5	28.46
branch, access, 67.5 degrees	nr	27.77	2.5	28.46
branch, access, 45 degrees	nr	27.77	2.5	28.46
branch, double, 87.5 degrees	nr	27.77	2.5	28.46
branch, double access, 87.5 degrees	nr	37.36	2.5	38.29
branch, parallel	nr	14.44	2.5	14.80
change piece, to soil	nr	7.93	2.5	8.13
offset, 75mm projection	nr	9.26	2.5	9.49
offset, 115mm projection	nr	11.52	2.5	11.81
offset, 150mm projection	nr	11.52	2.5	11.81
offset, 225mm projection	nr	14.92	2.5	15.29
offset, 300mm projection	nr	16.99	2.5	17.41
roof vent connector for asphalt	nr	24.11	2.5	24.71
roof vent connector for felt	nr	58.20	2.5	59.66
roof outlet	nr	31.45	2.5	32.24
wall fixing bracket	nr	5.10	2.5	5.23
100mm diameter pipe	m	14.99	5.0	15.74
bend, short radius	nr	12.94	2.5	13.26

Soil pipes (cont'd)	Unit	Material supply (£)	Material waste (%)	Total (£)
bend, short radius, access	nr	27.37	2.5	28.05
bend, long radius	nr	20.97	2.5	21.49
bend, long radius, access	nr	35.39	2.5	36.27
bend, long radius, rest	nr	25.40	2.5	26.04
branch, 87.5 degrees	nr	20.03	2.5	20.53
branch, 76 degrees	nr	20.03	2.5	20.53
branch, 67.5 degrees	nr	20.03	2.5	20.53
branch, 45 degrees	nr	20.03	2.5	20.53
branch, access, 87.5 degrees	nr	34.45	2.5	35.31
branch, access, 76 degrees	nr	34.45	2.5	35.31
branch, access, 67.5 degrees	nr	34.45	2.5	35.31
branch, access, 45 degrees	nr	34.45	2.5	35.31
branch, double, 87.5 degrees	nr	24.76	2.5	25.38
branch, double, 76 degrees	nr	24.76	2.5	25.38
branch, double, 67.5 degrees	nr	24.76	2.5	25.38
branch, double, 45 degrees	nr	24.76	2.5	25.38
branch, double access, 87.5 degrees	nr	39.20	2.5	40.18
branch, double access, 76 degrees	nr	39.20	2.5	40.18
branch, double access, 67.5 degrees	nr	39.20	2.5	40.18
branch, double access, 45 degrees	nr	39.20	2.5	40.18
branch, parallel	nr	20.97	2.5	21.49
change piece, to soil	nr	9.77	2.5	10.01
change piece, to drain	nr	25.40	2.5	26.04
change piece, to clay	nr	17.94	2.5	18.39
WC connector	nr	14.09	2.5	14.44
WC connecting bend, 300mm tail	nr	22.61	2.5	23.18
long tail bend, 87.5 degrees	nr	39.20	2.5	40.18
long tail bend, 76 degrees	nr	48.20	2.5	49.41
offset, 75mm projection	nr	13.60	2.5	13.94
offset, 115mm projection	nr	16.38	2.5	16.79
offset, 150mm projection	nr	16.23	2.5	16.64
offset, 225mm projection	nr	16.25	2.5	16.66
offset, 300mm projection	nr	20.97	2.5	21.49
roof vent connector for asphalt	nr	28.53	2.5	29.24
roof vent connector for felt	nr	70.83	2.5	72.60
roof outlet	nr	40.37	2.5	41.38
wall fixing bracket	nr	5.56	2.5	5.70

MATERIAL COSTS

	Unit	Material supply (£)	Material waste (%)	Total (£)
150mm diameter pipe	m	30.19	5.0	31.70
bend, short radius	nr	23.12	2.5	23.70
bend, short radius, access	nr	38.89	2.5	39.86
bend, long radius	nr	37.96	2.5	38.91
bend, long radius, access	nr	53.70	2.5	55.04
bend, long radius, rest	nr	43.44	2.5	44.53
branch, 87.5 degrees	nr	41.21	2.5	42.24
branch, 45 degrees	nr	41.21	2.5	42.24
branch, access, 87.5 degrees	nr	56.96	2.5	58.38
branch, access, 76 degrees	nr	53.72	2.5	55.06
branch, access, 45 degrees	nr	54.91	2.5	56.28
branch, double, 87.5 degrees	nr	63.23	2.5	64.81
branch, double, 76 degrees	nr	61.16	2.5	62.69
branch, double access, 87.5 degrees	nr	78.99	2.5	80.96
branch, double access, 76 degrees	nr	76.93	2.5	78.85
change piece, to soil	nr	19.45	2.5	19.94
change piece, to drain	nr	31.05	2.5	31.83
change piece, to clay	nr	20.19	2.5	20.69
offset, 75mm projection	nr	28.70	2.5	29.42
offset, 115mm projection	nr	33.64	2.5	34.48
offset, 150mm projection	nr	33.64	2.5	34.48
offset, 225mm projection	nr	38.22	2.5	39.18
offset, 300mm projection	nr	49.09	2.5	50.32
roof vent connector for asphalt	nr	52.99	2.5	54.31
roof vent connector for felt	nr	105.95	2.5	108.60
wall fixing bracket	nr	10.59	2.5	10.85

	Unit	Material supply (£)	Material waste (%)	Total (£)
OVERFLOWS				
uPVC, Terrain overflow system solvent-welded				
19mm diameter pipe	m	0.91	5.0	0.96
bend, 91 degrees	nr	0.84	2.5	0.86
bend, 135 degrees	nr	0.98	2.5	1.00
branch, 91 degrees	nr	0.98	2.5	1.00
connector, bent tank	nr	1.34	2.5	1.37
connector, straight tank	nr	1.19	2.5	1.22
connector, iron to uPVC	nr	1.10	2.5	1.13
pipe clip	nr	0.27	2.5	0.28
uPVC, Marley overflow system				
22mm diameter pipe	m	0.82	5.0	0.86
bend, 88.5 degrees	nr	0.81	2.5	0.83
bend, adjustable	nr	1.09	2.5	1.12
branch, 88.5 degrees	nr	0.74	2.5	0.76
adaptor, iron to uPVC	nr	0.68	2.5	0.70
adaptor, straight copper to uPVC	nr	0.55	2.5	0.56
straight tank connector	nr	1.16	2.5	1.19
bent tank connector	nr	1.23	2.5	1.26
saddle clip	nr	0.20	2.5	0.21
ABS, OsmaWeld system				
19mm diameter pipe	m	0.53	5.0	0.56
bend, 90 degrees	nr	0.57	2.5	0.58
tee, 90 degrees	nr	0.62	2.5	0.64
connector, bent tank	nr	0.84	2.5	0.86
pipe clip	nr	0.17	2.5	0.17

	Unit	Material supply (£)	Material waste (%)	Total (£)
Polypropylene, Osma/ClearBore, ring-seal socket joints				
19mm diameter pipe	m	0.41	5.0	0.43
bend, 90 degrees	nr	0.57	2.5	0.58
tee, 90 degrees	nr	0.62	2.5	0.64
connector, bent tank	nr	0.81	2.5	0.83
connector, straight tank	nr	0.81	2.5	0.83
pipe clip	nr	0.17	2.5	0.17
Polypropylene, Hepworth Push-Fit overflow system				
19mm diameter pipe	m	0.33	5.0	0.35
bend, 90 degrees	nr	0.42	2.5	0.43
tee, 90 degrees	nr	0.42	2.5	0.43
reducer	nr	0.42	2.5	0.43
connector, straight tank	nr	0.42	2.5	0.43
connector, bent tank	nr	0.42	2.5	0.43
pipe clip	nr	0.24	2.5	0.25
ABS, Hepworth overflow system				
19mm diameter pipe	m	0.33	5.0	0.35
bend, 90 degrees	nr	0.42	2.5	0.43
tee, 90 degrees	nr	0.42	2.5	0.43
reducer	nr	0.42	2.5	0.43
connector, straight tank	nr	0.42	2.5	0.43
connector, bent tank	nr	0.42	2.5	0.43
pipe clip	nr	0.24	2.5	0.25
TRAPS				
Polypropylene, Terrain, screwed joint to outlet and pipe				
bottle P trap				
32mm	nr	3.97	2.5	4.07
40mm	nr	4.75	2.5	4.87

Traps (cont'd)	Unit	Material supply (£)	Material waste (%)	Total (£)
bottle S trap				
32mm	nr	4.79	2.5	4.91
40mm	nr	5.82	2.5	5.97
self-resealing bottle P trap				
32mm	nr	4.95	2.5	5.07
40mm	nr	5.80	2.5	5.94
self-resealing bottle S trap				
32mm	nr	5.84	2.5	5.99
40mm	nr	6.77	2.5	6.94
telescopic bottle P trap				
32mm	nr	6.76	2.5	6.93
40mm	nr	8.13	2.5	8.33
tubular P trap				
32mm	nr	3.57	2.5	3.66
40mm	nr	4.12	2.5	4.22
50mm	nr	9.87	2.5	10.12
tubular S trap				
32mm	nr	4.53	2.5	4.64
40mm	nr	5.30	2.5	5.43
washing machine trap				
40mm	nr	7.16	2.5	7.34
tubular running P trap				
32mm	nr	5.48	2.5	5.62
40mm	nr	5.98	2.5	6.13
50mm	nr	13.96	2.5	14.31
tubular P trap for bath				
40mm	nr	7.66	2.5	7.85

MATERIAL COSTS

	Unit	Material supply (£)	Material waste (%)	Total (£)
tubular P trap assembly for bath				
40mm, white	nr	13.06	2.5	13.39
40mm, chromium-plated	nr	15.65	2.5	16.04
tubular P lowline bath trap				
40mm	nr	4.48	2.5	4.59
tubular P lowline bath trap with overflow				
40mm	nr	6.91	2.5	7.08
Polypropylene, Marley, screwed joint to outlet and pipe				
tubular P trap				
32mm	nr	2.57	2.5	2.63
40mm	nr	3.08	2.5	3.16
tubular S trap				
32mm	nr	3.08	2.5	3.16
40mm	nr	3.67	2.5	3.76
standard bottle trap				
32mm	nr	2.63	2.5	2.70
40mm	nr	2.47	2.5	2.53
monitor anti-syphon bottle trap				
32mm	nr	4.18	2.5	4.28
40mm	nr	4.36	2.5	4.47
low inlet tubular P trap, 40mm				
plain	nr	3.08	2.5	3.16
with access	nr	3.60	2.5	3.69
with overflow and outlet	nr	8.89	2.5	9.11

Traps (cont'd)	Unit	Material supply (£)	Material waste (%)	Total (£)
shallow bath trap, 40mm	nr	3.88	2.5	3.98
low level bath P trap	nr	3.63	2.5	3.72
washing machine kit	nr	5.20	2.5	5.33
running P trap				
40mm	nr	4.42	2.5	4.53
50mm	nr	10.45	2.5	10.71
Polypropylene, Osma, screwed joint to outlet and pipe				
bottle trap with 38mm seal				
32mm	nr	2.17	2.5	2.22
40mm	nr	2.70	2.5	2.77
bottle trap with 76mm seal				
32mm	nr	2.50	2.5	2.56
40mm	nr	2.97	2.5	3.04
tubular P trap				
32mm	nr	2.31	2.5	2.37
40mm	nr	2.78	2.5	2.85
tubular S trap				
32mm	nr	2.77	2.5	2.84
40mm	nr	3.37	2.5	3.45
adjustable tubular P trap				
32mm	nr	2.82	2.5	2.89
40mm	nr	3.32	2.5	3.40
running P trap				
32mm	nr	3.48	2.5	3.57
40mm	nr	3.80	2.5	3.90

	Unit	Material supply (£)	Material waste (%)	Total (£)
bath trap, shallow seal				
32mm	nr	2.06	2.5	2.11
40mm	nr	2.22	2.5	2.28
bath trap, 76mm seal				
40mm	nr	3.12	2.5	3.20
bath trap and overflow				
40mm	nr	7.05	2.5	7.23
washing machine half trap				
40mm	nr	5.07	2.5	5.20
washing machine trap				
40mm	nr	4.38	2.5	4.49
Polypropylene, Hepworth, screwed joint to outlet and pipe				
bottle trap with 38mm seal				
32mm	nr	1.85	2.5	1.90
40mm	nr	2.14	2.5	2.19
bottle trap with 76mm seal				
32mm	nr	1.78	2.5	1.82
40mm	nr	2.34	2.5	2.40
tubular P trap				
32mm	nr	1.64	2.5	1.68
40mm	nr	1.91	2.5	1.96
tubular S trap				
32mm	nr	2.31	2.5	2.37
40mm	nr	2.70	2.5	2.77

Traps (cont'd)	Unit	Material supply (£)	Material waste (%)	Total (£)
adjustable tubular P trap				
32mm	nr	2.05	2.5	2.10
40mm	nr	2.17	2.5	2.22
running P trap				
40mm	nr	3.26	2.5	3.34
washing machine trap				
40mm	nr	3.69	2.5	3.78

	Unit	Labour hours	Net labour (£)	Net material (£)	O'heads /profit (£)	Total (£)
WASTE PIPES						
MuPVC Terrain waste system 200, solvent-welded joints, to brickwork with clips at 500mm maximum centres, plugging						
32mm diameter pipe	m	0.24	2.30	2.43	0.71	5.44
Extra over for						
bend, 91.25 degrees	nr	0.22	2.11	1.81	0.59	4.51
bend, 135 degrees	nr	0.22	2.11	1.81	0.59	4.51
bend, 165 degrees	nr	0.22	2.11	1.93	0.61	4.65
tee, 91.25 degrees	nr	0.25	2.40	2.58	0.75	5.73
tee, 135 degrees	nr	0.25	2.40	2.78	0.78	5.96
connector, iron to MuPVC	nr	0.22	2.11	1.55	0.55	4.21
connector, copper to MuPVC	nr	0.22	2.11	2.05	0.62	4.78
connection to back inlet gulley, caulking bush	nr	0.15	1.44	2.16	0.54	4.14
0mm diameter pipe	m	0.27	2.59	2.99	0.84	6.42
Extra over for						
bend, 91.25 degrees	nr	0.24	2.30	2.05	0.65	5.00
bend, 135 degrees	nr	0.24	2.30	2.05	0.65	5.00
bend, 165 degrees	nr	0.24	2.30	2.27	0.69	5.26
tee, 91.25 degrees	nr	0.28	2.69	3.22	0.89	6.80
tee, 135 degrees	nr	0.28	2.69	3.36	0.91	6.96
sweep cross	nr	0.32	3.07	9.26	1.85	14.18
connector, iron to MuPVC	nr	0.24	2.30	1.82	0.62	4.74
connector, copper to MuPVC	nr	0.24	2.30	2.16	0.67	5.13
connection to back inlet gulley, caulking bush	nr	0.17	1.63	2.16	0.57	4.36
0mm diameter pipe	m	0.30	2.88	4.59	1.12	8.59
Extra over for						
bend, 91.25 degrees	nr	0.26	2.50	2.95	0.82	6.27
bend, 135 degrees	nr	0.26	2.50	3.03	0.83	6.36
bend, 165 degrees	nr	0.26	2.50	3.21	0.86	6.57
tee, 91.25 degrees	nr	0.30	2.88	5.62	1.27	9.78
tee, 135 degrees	nr	0.30	2.88	5.86	1.31	10.05
sweep cross	nr	0.32	3.07	9.36	1.86	14.29
connector, iron to MuPVC	nr	0.26	2.50	2.61	0.77	5.88

Waste pipes (cont'd)	Unit	Labour hours	Net labour (£)	Net material (£)	O'heads /profit (£)	Total (£)
connector, copper to MuPVC	nr	0.26	2.50	2.16	0.70	5.36
connection to back inlet gulley, caulking bush	nr	0.20	1.92	1.99	0.59	4.50
MuPVC, Marley waste system, solvent-welded joints, to brickwork with clips at 500mm maximum centres, plugging						
32mm diameter pipe	m	0.24	2.30	2.06	0.65	5.01
Extra over for						
branch, 88.5 degrees	nr	0.25	2.40	1.56	0.59	4.55
branch, 45 degrees	nr	0.25	2.40	1.56	0.59	4.55
reducer, 32 to 22mm	nr	0.22	2.11	1.31	0.51	3.93
adaptor, iron to MuPVC	nr	0.22	2.11	1.56	0.55	4.22
40mm diameter pipe	m	0.27	2.59	2.50	0.76	5.85
Extra over for						
branch, 88.5 degrees	nr	0.28	2.69	1.76	0.67	5.12
branch, 45 degrees	nr	0.28	2.69	1.76	0.67	5.12
reducer, 40 to 32mm	nr	0.24	2.30	1.09	0.51	3.90
cross tee	nr	0.30	2.88	8.49	1.71	13.08
adaptor, iron to MuPVC	nr	0.24	2.30	1.76	0.61	4.67
50mm diameter pipe	m	0.30	2.88	3.76	1.00	7.64
Extra over for						
branch, 88.5 degrees	nr	0.30	2.88	2.47	0.80	6.1
branch, 45 degrees	nr	0.30	2.88	2.47	0.80	6.1
cross tee	nr	0.34	3.26	9.85	1.97	15.0
reducer, 50 to 40mm	nr	0.26	2.50	1.82	0.65	4.9
reducer, 50 to 32mm	nr	0.26	2.50	2.01	0.68	5.1
adaptor, iron to MuPVC	nr	0.26	2.50	2.62	0.77	5.8
connection to back inlet gulley, caulking bush	nr	0.20	1.92	2.84	0.71	5.4

	Unit	Labour hours	Net labour (£)	Net material (£)	O'heads /profit (£)	Total (£)
Ultra ABS, Marley waste system, solvent-welded joints, to clips at 500mm maximum centres, plugging						
32mm diameter pipe	m	0.24	2.30	1.34	0.55	4.19
Extra over for						
bend, 45 degrees	nr	0.22	2.11	0.99	0.46	3.57
bend, 88.5 degrees	nr	0.22	2.11	1.16	0.49	3.76
branch, 88.5 degrees	nr	0.24	2.30	1.71	0.60	4.61
branch, 45 degrees	nr	0.24	2.30	1.71	0.60	4.61
adaptor, iron to MuPVC	nr	0.22	2.11	1.37	0.52	4.00
40mm diameter pipe	m	0.27	2.59	1.67	0.64	4.90
Extra over for						
bend, 88.5 degrees	nr	0.24	2.30	1.37	0.55	4.22
bend, 45 degrees	nr	0.24	2.30	1.16	0.52	3.98
branch, 88.5 degrees	nr	0.28	2.69	1.98	0.70	5.37
branch, 45 degrees	nr	0.28	2.69	1.98	0.70	5.37
cross tee	nr	0.30	2.88	3.60	0.97	7.45
reducer, 40 to 32mm	nr	0.24	2.30	0.43	0.41	3.14
adaptor, iron to MuPVC	nr	0.24	2.30	1.47	0.57	4.34
50mm diameter pipe	m	0.30	2.88	2.29	0.78	5.95
Extra over for						
bend, 88.5 degrees	nr	0.26	2.50	2.19	0.70	5.39
bend, 45 degrees	nr	0.26	2.50	1.98	0.67	5.15
branch, 88.5 degrees	nr	0.30	2.88	3.00	0.88	6.76
branch, 45 degrees	nr	0.30	2.88	3.00	0.88	6.76
cross tee	nr	0.34	3.26	4.91	1.23	9.40
reducer, 50 to 40mm	nr	0.26	2.50	0.97	0.52	3.99
reducer, 50 to 32 mm	nr	0.26	2.50	0.97	0.52	3.99
adaptor, iron to MuPVC	nr	0.26	2.50	1.98	0.67	5.15
connection to back inlet gulley, caulking bush	nr	0.20	1.92	2.84	0.71	5.47

Waste pipes (cont'd)	Unit	Labour hours	Net labour (£)	Net material (£)	O'heads /profit (£)	Total (£)
MuPVC, Marley waste system, push-fit joints, to brickwork with clips at 500mm maximum centres, plugging						
32mm diameter pipe	m	0.21	2.02	0.74	0.41	3.17
Extra over for						
bend, 90 degrees	nr	0.19	1.82	0.76	0.39	2.97
bend, 45 degrees	nr	0.19	1.82	0.79	0.39	3.00
branch, 88.5 degrees	nr	0.22	2.11	0.98	0.46	3.55
tank connector	nr	0.20	1.92	1.16	0.46	3.54
multi-fit waste connector	nr	0.20	1.92	1.49	0.51	3.92
40mm diameter pipe	m	0.21	2.02	0.91	0.44	3.37
Extra over for						
bend, 90 degrees	nr	0.19	1.82	0.79	0.39	3.00
bend, 45 degrees	nr	0.19	1.82	0.85	0.40	3.07
branch, 88.5 degrees	nr	0.22	2.11	1.27	0.51	3.89
reducer, 40 to 32mm	nr	0.22	2.11	0.85	0.44	3.40
tank connector	nr	0.20	1.92	1.23	0.47	3.62
multi-fit waste connector	nr	0.20	1.92	1.54	0.52	3.98
ABS, OsmaWeld waste system, solvent-welded joints, to brickwork with clips at 500mm maximum centres, plugging						
32mm diameter pipe	m	0.24	2.30	1.45	0.56	4.3
Extra over for						
bend, 90 degrees	nr	0.22	2.11	0.94	0.46	3.5
bend, 87.5 degrees	nr	0.22	2.11	0.99	0.46	3.5
bend, 45 degrees	nr	0.22	2.11	0.88	0.45	3.4
bend, long tail 90 degrees	nr	0.22	2.11	1.06	0.48	3.6
tee	nr	0.25	2.40	1.41	0.57	4.3
access plug	nr	0.10	0.96	0.82	0.27	2.0
connector, straight tank	nr	0.22	2.11	1.77	0.58	4.4
connector, universal	nr	0.22	2.11	1.27	0.51	3.8
connector, copper to ABS	nr	0.22	2.11	1.77	0.58	4.4
connector, iron to ABS	nr	0.22	2.11	1.50	0.54	4.1
reducer, 32 to 19mm	nr	0.22	2.11	0.70	0.42	3.2

	Unit	Labour hours	Net labour (£)	Net material (£)	O'heads /profit (£)	Total (£)
40mm diameter pipe	m	0.27	2.59	1.55	0.62	4.76
Extra over for						
bend, 90 degrees	nr	0.24	2.30	1.13	0.51	3.94
bend, 87.5 degrees	nr	0.24	2.30	1.20	0.53	4.03
bend, 45 degrees	nr	0.24	2.30	0.99	0.49	3.78
bend, long tail, 90 degrees	nr	0.24	2.30	1.26	0.53	4.09
tee	nr	0.28	2.69	1.73	0.66	5.08
cross tee	nr	0.30	2.88	3.29	0.93	7.10
access plug	nr	0.10	0.96	0.85	0.27	2.08
connector, straight tank	nr	0.24	2.30	1.94	0.64	4.88
connector, universal	nr	0.24	2.30	1.37	0.55	4.22
connector, copper to ABS	nr	0.24	2.30	1.94	0.64	4.88
connector, iron to ABS	nr	0.24	2.30	1.58	0.58	4.46
reducer, 40 to 32mm	nr	0.24	2.30	0.79	0.46	3.55
50mm diameter pipe	m	0.30	2.88	2.40	0.79	6.07
Extra over for						
bend, 87.5 degrees	nr	0.26	2.50	1.80	0.64	4.94
bend, 45 degrees	nr	0.26	2.50	1.57	0.61	4.68
bend, long tail, 90 degrees	nr	0.26	2.50	2.24	0.71	5.45
tee	nr	0.30	2.88	2.50	0.81	6.19
cross tee	nr	0.32	3.07	4.60	1.15	8.82
access plug	nr	0.10	0.96	1.41	0.36	2.73
connector, iron to ABS	nr	0.26	2.50	1.97	0.67	5.14
reducer, 50 to 40mm	nr	0.26	2.50	1.03	0.53	4.06
reducer, 50 to 32mm	nr	0.26	2.50	1.03	0.53	4.06

Polypropylene, Osma ClearBore ing-seal joints, to brickwork with clips at 500mm maximum centres, plugging

	Unit	Labour hours	Net labour (£)	Net material (£)	O'heads /profit (£)	Total (£)
2mm diameter pipe	m	0.21	2.02	1.21	0.48	3.71
Extra over for						
bend, 90 degrees	nr	0.19	1.82	1.23	0.46	3.51
bend, 87.5 degrees	nr	0.19	1.82	1.25	0.46	3.53
bend, 45 degrees	nr	0.19	1.82	1.14	0.44	3.40
bend, spigot, 45 degrees	nr	0.19	1.82	1.09	0.44	3.35
tee	nr	0.22	2.11	1.78	0.58	4.47
access plug	nr	0.10	0.96	0.82	0.27	2.05
connector, straight tank	nr	0.19	1.82	1.77	0.54	4.13

Waste pipes (cont'd)	Unit	Labour hours	Net labour (£)	Net material (£)	O'heads /profit (£)	Total (£)
connector, universal	nr	0.19	1.82	1.27	0.46	3.55
connector, copper to ABS	nr	0.19	1.82	1.57	0.51	3.90
connector, iron to ABS	nr	0.19	1.82	1.50	0.50	3.82
reducer, 32 to 19mm	nr	0.19	1.82	0.70	0.38	2.90
40mm diameter pipe	m	0.24	2.30	1.43	0.56	4.29
Extra over for						
bend, 90 degrees	nr	0.21	2.02	1.33	0.50	3.85
bend, 87.5 degrees	nr	0.21	2.02	1.40	0.51	3.93
bend, 45 degrees	nr	0.21	2.02	1.20	0.48	3.70
bend, spigot, 45 degrees	nr	0.21	2.02	1.14	0.47	3.63
tee	nr	0.25	2.40	2.06	0.67	5.13
access plug	nr	0.10	0.96	0.85	0.27	2.08
connector, straight tank	nr	0.21	2.02	1.94	0.59	4.55
connector, universal	nr	0.21	2.02	2.01	0.60	4.63
connector, copper to ABS	nr	0.21	2.02	1.78	0.57	4.37
connector, iron to ABS	nr	0.21	2.02	1.58	0.54	4.14
reducer, 40 to 32mm	nr	0.21	2.02	0.78	0.42	3.22
50mm diameter pipe	m	0.27	2.59	2.30	0.73	5.6[illegible]
Extra over for						
bend, 87.5 degrees	nr	0.23	2.21	2.46	0.70	5.3[illegible]
bend, 45 degrees	nr	0.23	2.21	2.18	0.66	5.0[illegible]
tee	nr	0.27	2.59	3.00	0.84	6.4[illegible]
access plug	nr	0.10	0.96	1.41	0.36	2.7[illegible]
connector, iron to ABS	nr	0.23	2.21	1.13	0.50	3.8[illegible]
reducer, 50 to 40mm	nr	0.23	2.21	1.16	0.51	3.8[illegible]
reducer, 50 to 32mm	nr	0.23	2.21	1.16	0.51	3.8[illegible]
Polypropylene, Hepworth waste systems, ring-seal connector joints, to brickwork with 500mm maximum centres, plugging						
32mm diameter pipe	m	0.21	2.02	0.92	0.44	3.3
Extra over for						
reducer, 32 to 21.5mm	nr	0.19	1.82	0.44	0.34	2.6
tank connector, to iron	nr	0.19	1.82	0.58	0.36	2.7
tank connector, to copper	nr	0.19	1.82	0.58	0.36	2.7
bend, 135 degrees	nr	0.19	1.82	0.58	0.36	2.7
bend, 150 degrees	nr	0.19	1.82	0.58	0.36	2.7

	Unit	Labour hours	Net labour (£)	Net material (£)	O'heads /profit (£)	Total (£)
bend, 90 degrees	nr	0. 19	1. 82	0. 58	0. 36	2. 76
bend, 92.5 degrees	nr	0. 19	1. 82	0. 58	0. 36	2. 76
tee, 92.5 degrees	nr	0. 22	2. 11	0. 58	0. 40	3. 09
plug, access	nr	0. 10	0. 96	0. 58	0. 23	1. 77
40mm diameter pipe	m	0. 24	2. 30	1. 04	0. 50	3. 84
Extra over for						
reducer, 40 to 21.5mm	nr	0. 21	2. 02	0. 44	0. 37	2. 83
reducer, 40 to 32mm	nr	0. 21	2. 02	0. 58	0. 39	2. 99
tank connector, to iron	nr	0. 21	2. 02	0. 58	0. 39	2. 99
tank connector, to copper	nr	0. 21	2. 02	0. 58	0. 39	2. 99
bend, 135 degrees	nr	0. 21	2. 02	0. 58	0. 39	2. 99
bend, 150 degrees	nr	0. 21	2. 02	0. 58	0. 39	2. 99
bend, 90 degrees	nr	0. 21	2. 02	0. 58	0. 39	2. 99
bend, 92.5 degrees	nr	0. 21	2. 02	0. 58	0. 39	2. 99
tee, 92.5 degrees	nr	0. 25	2. 40	0. 58	0. 45	3. 43
plug, access	nr	0. 10	0. 96	0. 58	0. 23	1. 77
50mm diameter pipe	m	0. 27	2. 59	1. 78	0. 66	5. 03
Extra over for						
reducer, 50 to 40mm	nr	0. 23	2. 21	1. 06	0. 49	3. 76
reducer, 50 to 32mm	nr	0. 23	2. 21	1. 06	0. 49	3. 76
bend, 135 degrees	nr	0. 23	2. 21	1. 06	0. 49	3. 76
bend, 150 degrees	nr	0. 23	2. 21	1. 06	0. 49	3. 76
bend, 90 degrees	nr	0. 23	2. 21	1. 06	0. 49	3. 76
bend, 92.5 degrees	nr	0. 23	2. 21	1. 06	0. 49	3. 76
tee, 92.5 degrees	nr	0. 27	2. 59	1. 06	0. 55	4. 20
plug, access	nr	0. 10	0. 96	1. 06	0. 30	2. 32

Waste pipes (cont'd)	Unit	Labour hours	Net labour (£)	Net material (£)	O'heads /profit (£)	Total (£)
ABS, Hepworth waste system, solvent-welded joints, connector joints, to brickwork at 500mm maximum centres, plugging						
32mm diameter pipe	m	0.24	2.30	1.38	0.55	4.23
Extra over for						
bend, 135 degrees	nr	0.22	2.11	0.66	0.42	3.19
bend, 90 degrees	nr	0.22	2.11	0.66	0.42	3.19
bend, 92.5 degrees	nr	0.22	2.11	0.66	0.42	3.19
tee, 92.5 degrees	nr	0.25	2.40	0.66	0.46	3.52
plug, access	nr	0.10	0.96	0.66	0.24	1.86
40mm diameter pipe	m	0.27	2.59	1.62	0.63	4.84
Extra over for						
reducer, 40 to 32mm	nr	0.24	2.30	0.66	0.44	3.40
bend, 135 degrees	nr	0.24	2.30	0.66	0.44	3.40
bend, 90 degrees	nr	0.24	2.30	0.66	0.44	3.40
bend, 92.5 degrees	nr	0.24	2.30	0.66	0.44	3.40
tee, 92.5 degrees	nr	0.28	2.69	0.66	0.50	3.85
tee, cross	nr	0.30	2.88	0.66	0.53	4.07
plug, access	nr	0.10	0.96	0.66	0.24	1.86
50mm diameter pipe	m	1.30	12.48	2.09	2.19	16.76
Extra over for						
reducer, 50 to 40mm	nr	0.26	2.50	0.89	0.51	3.90
reducer, 50 to 32mm	nr	0.26	2.50	0.89	0.51	3.90
bend, 135 degrees	nr	0.26	2.50	0.89	0.51	3.90
bend, 90 degrees	nr	0.26	2.50	0.89	0.51	3.90
bend, 92.5 degrees	nr	0.26	2.50	0.89	0.51	3.90
tee, 92.5 degrees	nr	0.30	2.88	0.89	0.57	4.34
tee, cross	nr	0.34	3.26	0.89	0.62	4.77
plug, access	nr	0.10	0.96	0.89	0.28	2.13

	Unit	Labour hours	Net labour (£)	Net material (£)	O'heads /profit (£)	Total (£)
SOIL PIPES						
uPVC, Terrain soil system 100, solvent-welded joints, to brickwork with holderbats at 1250mm maximum centres, plugging						
82mm diameter pipe	m	0.33	3.17	6.95	1.52	11.64
Extra over for						
bend, 92.5 degrees	nr	0.30	2.88	7.48	1.55	11.91
bend, 135 degrees	nr	0.30	2.88	7.48	1.55	11.91
branch, single, 92.5 degrees	nr	0.35	3.36	10.40	2.06	15.82
branch single, 135 degrees	nr	0.35	3.36	10.40	2.06	15.82
vent cowl	nr	0.20	1.92	2.09	0.60	4.61
weathered apron	nr	0.22	2.11	2.09	0.63	4.83
110mm diameter pipe	m	0.37	3.55	8.17	1.76	13.48
Extra over for						
bend, 92.5 degrees	nr	0.34	3.26	10.24	2.02	15.53
bend, 135 degrees	nr	0.34	3.26	10.24	2.02	15.53
bend, variable, spigot/spigot	nr	0.34	3.26	14.59	2.68	20.53
bend, variable, spigot/socket	nr	0.34	3.26	14.59	2.68	20.53
bend, WC connecting, 92.5 degrees	nr	0.34	3.26	10.91	2.13	16.30
bend, access, 92.5 degrees	nr	0.34	3.26	22.98	3.94	30.18
branch, single, 92.5 degrees	nr	0.39	3.74	13.40	2.57	19.71
branch, single, 104 degrees	nr	0.39	3.74	13.55	2.59	19.88
branch, single, 135 degrees	nr	0.39	3.74	13.55	2.59	19.88
branch, single, access	nr	0.39	3.74	26.24	4.50	34.48
branch, double, 92.5 degrees	nr	0.39	3.74	31.59	5.30	40.63
branch, double, 104 degrees	nr	0.39	3.74	31.59	5.30	40.63
branch, double, 135 degrees	nr	0.39	3.74	33.03	5.52	42.29
branch, corner, 92.5 degrees	nr	0.39	3.74	57.15	9.13	70.02
bossed pipe connector, solvent welded sockets, single	nr	0.34	3.26	3.26	0.98	7.50
bossed pipe connector, spigot boss and solvent welded sockets	nr	0.34	3.26	3.38	1.00	7.64
connector, WC horizontal outlet	nr	0.30	2.88	9.25	1.82	13.95
connector, WC variable	nr	0.30	2.88	15.57	2.77	21.22
vent cowl	nr	0.22	2.11	2.11	0.63	4.85
weathered apron	nr	0.24	2.30	2.46	0.71	5.47

Soil pipes (cont'd)	Unit	Labour hours	Net labour (£)	Net material (£)	O'heads /profit (£)	Total (£)
160mm diameter pipe	m	0.42	4.03	17.75	3.27	25.05
Extra over for						
bend, 92.5 degrees	nr	0.37	3.55	22.08	3.84	29.47
bend, 135 degrees	nr	0.37	3.55	20.90	3.67	28.12
branch, single, 92.5 degrees	nr	0.48	4.61	70.54	11.27	86.42
branch, single, 135 degrees	nr	0.48	4.61	45.32	7.49	57.42
branch, double, 92.5 degrees	nr	0.48	4.61	49.33	8.09	62.03
branch, double, 104 degrees	nr	0.48	4.61	55.04	8.95	68.60
vent cowl	nr	0.24	2.30	5.52	1.17	8.99
weathered apron	nr	0.26	2.50	7.43	1.49	11.42
uPVC, Marley soil system, spigot and socket seal joints, to brickwork with brackets at 1250mm maximum centres, plugging						
82mm diameter pipe	m	0.30	2.88	6.66	1.43	10.97
Extra over for						
bend, spigot/socket	nr	0.27	2.59	7.38	1.50	11.47
bend, solvent socket	nr	0.27	2.59	7.38	1.50	11.47
bend, adjustable	nr	0.27	2.59	12.36	2.24	17.19
offset bend, socket	nr	0.27	2.59	7.38	1.50	11.47
offset bend, spigot	nr	0.27	2.59	6.23	1.32	10.14
branch, socket/socket/solvent socket, plain	nr	0.31	2.98	10.81	2.07	15.86
branch, adjustable socket/ spigot, 3 boss upstand	nr	0.31	2.98	11.57	2.18	16.73
boss branch, 3 boss upstand	nr	0.31	2.98	5.48	1.27	9.73
boss branch, single 50mm, 3 boss upstand/solvent sockets	nr	0.31	2.98	5.87	1.33	10.18
rear access bend, adjustable	nr	0.27	2.59	19.42	3.30	25.31
rear access branch, adjustable	nr	0.31	2.98	20.43	3.51	26.92
vent terminal	nr	0.20	1.92	1.99	0.59	4.50
vent slate cap	nr	0.20	1.92	2.69	0.69	5.30
WC connector, straight	nr	0.27	2.59	3.22	0.87	6.68
110mm diameter pipe	m	0.34	3.26	6.97	1.53	11.76
Extra over for						
bend, spigot/socket	nr	0.31	2.98	9.75	1.91	14.64
bend, solvent socket	nr	0.31	2.98	9.75	1.91	14.64
bend, adjustable	nr	0.31	2.98	12.36	2.30	17.64

	Unit	Labour hours	Net labour (£)	Net material (£)	O'heads /profit (£)	Total (£)
offset bend, end socket	nr	0.31	2.98	8.20	1.68	12.86
branch, single socket/spigot, plain	nr	0.36	3.46	15.02	2.77	21.25
branch, single socket/spigot, 5 boss upstand	nr	0.36	3.46	12.69	2.42	18.57
branch, socket/socket/solvent socket, 3 boss upstand	nr	0.36	3.46	15.02	2.77	21.25
branch, socket/socket/solvent socket, plain	nr	0.36	3.46	15.02	2.77	21.25
branch, double socket/spigot, corner	nr	0.36	3.46	52.24	8.36	64.06
branch, double socket/spigot, 2 boss upstand/2 solvent sockets	nr	0.36	3.46	28.98	4.87	37.31
multi-branch	nr	0.36	3.46	35.97	5.91	45.34
multi-branch, socket	nr	0.36	3.46	4.44	1.19	9.09
multi-branch, access cap	nr	0.36	3.46	7.94	1.71	13.11
boss branch, collar boss	nr	0.36	3.46	15.43	2.83	21.72
boss branch, single 40mm	nr	0.36	3.46	6.74	1.53	11.73
boss branch, single 32mm	nr	0.36	3.46	6.74	1.53	11.73
boss branch, single 50mm, all solvent connections	nr	0.36	3.46	3.16	0.99	7.61
boss branch, single 40mm, double solvent sockets	nr	0.36	3.46	2.93	0.96	7.35
boss branch, single 32mm, double solvent sockets	nr	0.36	3.46	2.93	0.96	7.35
boss branch, 4 boss upstand	nr	0.36	3.46	6.41	1.48	11.35
boss branch, single 50mm, 3 boss upstand/solvent sockets	nr	0.36	3.46	5.85	1.40	10.71
rear access bend, plain	nr	0.31	2.98	22.71	3.85	29.54
rear access branch, 2 boss upstand	nr	0.36	3.46	26.74	4.53	34.73
roof cowl	nr	0.20	1.92	11.40	2.00	15.32
vent terminal	nr	0.20	1.92	2.20	0.62	4.74
weathering collar	nr	0.25	2.40	2.20	0.69	5.29
WC connector, straight	nr	0.31	2.98	3.42	0.96	7.36
adaptor, to salt glazed socket	nr	0.25	2.40	5.18	1.14	8.72
adaptor, to cast iron socket	nr	0.25	2.40	12.56	2.24	17.20
160mm diameter pipe	m	0.37	3.55	16.51	3.01	23.07
Extra over for						
bend, spigot/socket	nr	0.34	3.26	23.29	3.98	30.53
bend, solvent socket	nr	0.34	3.26	23.29	3.98	30.53

Soil pipes (cont'd)	Unit	Labour hours	Net labour (£)	Net material (£)	O'heads /profit (£)	Total (£)
bend, adjustable	nr	0.34	3.26	23.48	4.01	30.75
offset bend, socket	nr	0.34	3.26	23.96	4.08	31.30
offset bend, spigot	nr	0.34	3.26	21.97	3.78	29.01
branch, single socket/spigot, plain	nr	0.37	3.55	38.37	6.29	48.21
branch, triple solvent socket, 3 boss upstand	nr	0.37	3.55	28.00	4.73	36.28
branch, socket/socket/solvent socket, plain	nr	0.37	3.55	30.61	5.12	39.28
rear access bend, adjustable	nr	0.34	3.26	35.58	5.83	44.67
vent terminal	nr	0.20	1.92	5.81	1.16	8.89
vent slate cap	nr	0.20	1.92	10.23	1.82	13.97
adaptor, to cast iron socket	nr	0.25	2.40	33.96	5.45	41.81
uPVC, Osmasoil system spigot and socket ring-seal joints, to brickwork at 1250mm maximum centres, plugging						
82mm diameter pipe	m	0.30	2.88	5.10	1.20	9.18
Extra over for						
reducer, 82 to 55mm	nr	0.27	2.59	5.19	1.17	8.95
connector, uPVC to cast iron	nr	0.24	2.30	4.91	1.08	8.29
bend, 87.5, 67.5, 45 degrees	nr	0.27	2.59	6.25	1.33	10.17
branch 87.5 degrees	nr	0.27	2.59	9.68	1.84	14.11
branch, 67.5 degrees	nr	0.31	2.98	9.68	1.90	14.56
weathering collar	nr	0.20	1.92	2.05	0.60	4.57
balloon grating	nr	0.15	1.44	1.78	0.48	3.70
110mm diameter pipe	m	0.34	3.26	5.75	1.35	10.36
Extra over for						
reducer, 110 to 82mm	nr	0.31	2.98	5.57	1.28	9.83
reducer, 110 to 55mm	nr	0.31	2.98	6.41	1.41	10.80
connector, uPVC to cast iron	nr	0.22	2.11	6.44	1.28	9.83
connector, uPVC to clay	nr	0.22	2.11	3.66	0.87	6.64
bend, 87.5, 67.5, 45 degrees	nr	0.31	2.98	8.38	1.70	13.06
bend, 30 or 15 degrees	nr	0.31	2.98	8.07	1.66	12.71
bend, adjustable, 30 degrees	nr	0.31	2.98	9.21	1.83	14.02
branch 87.5 degrees	nr	0.36	3.46	11.83	2.29	17.58
branch, 5 boss, 87.5 degrees	nr	0.36	3.46	12.30	2.36	18.12
branch, 67.5 degrees	nr	0.36	3.46	12.72	2.43	18.61
branch, 45 degrees	nr	0.36	3.46	12.34	2.37	18.17

	Unit	Labour hours	Net labour (£)	Net material (£)	O'heads /profit (£)	Total (£)
branch, 76 degrees	nr	0.36	3.46	12.72	2.43	18.61
branch, double, 45 degrees	nr	0.36	3.46	27.36	4.62	35.44
branch, corner, 87.5 degrees	nr	0.36	3.46	46.62	7.51	57.59
connector, WC, 90 degrees	nr	0.32	3.07	7.98	1.66	12.71
weathering collar	nr	0.15	1.44	2.02	0.52	3.98
balloon grating	nr	0.15	1.44	1.64	0.46	3.54
vent cowl	nr	0.20	1.92	9.07	1.65	12.64
160mm diameter pipe	m	0.37	3.55	13.71	2.59	19.85
Extra over for						
reducer, 160 to 110mm	nr	0.34	3.26	11.41	2.20	16.87
connector, uPVC to cast iron	nr	0.24	2.30	13.19	2.32	17.81
bend, 87.5, 67.5, 45 degrees	nr	0.34	3.26	19.75	3.45	26.46
bend, offset, bottom	nr	0.34	3.26	20.26	3.53	27.05
bend, offset, top	nr	0.34	3.26	22.31	3.84	29.41
branch, equal, 87.5 degrees	nr	0.40	3.84	35.66	5.92	45.43
branch, unequal, 87.5 degrees	nr	0.40	3.84	27.49	4.70	36.03
branch, unequal double, 87.5 degrees	nr	0.40	3.84	49.16	7.95	60.95
weathering collar	nr	0.22	2.11	6.75	1.33	10.19
balloon grating	nr	0.15	1.44	5.11	0.98	7.53
Cast iron to BS416, Drainage System Ltd 'Timesaver' system flexible joints, to brickwork with pipe brackets 2000mm maximum centres, plugging						
50mm diameter pipe	m	0.40	3.84	15.68	2.93	22.45
Extra over for						
bend, short radius	nr	0.30	2.88	9.82	1.90	14.60
bend, short radius, access	nr	0.30	2.88	24.21	4.06	31.15
branch, 87.5 degrees	nr	0.35	3.36	14.77	2.72	20.85
branch, 67.5 degrees	nr	0.35	3.36	14.78	2.72	20.86
branch, 45 degrees	nr	0.35	3.36	14.78	2.72	20.86
branch, access, 87.5 degrees	nr	0.35	3.36	29.17	4.88	37.41
branch, access, 67.5 degrees	nr	0.35	3.36	29.17	4.88	37.41
branch, access, 45 degrees	nr	0.35	3.36	29.17	4.88	37.41
branch, double, 87.5 degrees	nr	0.35	3.36	24.87	4.23	32.46
branch, double access, 87.5 degrees	nr	0.35	3.36	39.25	6.39	49.00
branch, parallel	nr	0.35	3.36	13.39	2.51	19.26

Soil pipes (cont'd)	Unit	Labour hours	Net labour (£)	Net material (£)	O'heads /profit (£)	Total (£)
change piece, to soil	nr	0.35	3.36	8.33	1.75	13.44
offset, 75mm projection	nr	0.35	3.36	9.73	1.96	15.05
offset, 115mm projection	nr	0.35	3.36	12.11	2.32	17.79
offset, 150mm projection	nr	0.35	3.36	15.67	2.85	21.88
offset, 225mm projection	nr	0.35	3.36	15.67	2.85	21.88
offset, 300mm projection	nr	0.35	3.36	17.85	3.18	24.39
roof vent connector for asphalt	nr	0.25	2.40	25.33	4.16	31.89
roof vent connector for felt	nr	0.00	0.00	61.15	9.17	70.32
roof outlet	nr	0.25	2.40	32.82	5.28	40.50
75mm diameter pipe	m	0.45	4.32	15.74	3.01	23.07
Extra over for						
bend, short radius	nr	0.40	3.84	9.82	2.05	15.71
bend, short radius, access	nr	0.40	3.84	23.47	4.10	31.41
bend, long radius	nr	0.40	3.84	18.54	3.36	25.74
bend, long radius, access	nr	0.40	3.84	32.93	5.52	42.29
branch, 87.5 degrees	nr	0.45	4.32	14.78	2.87	21.97
branch, 67.5 degrees	nr	0.45	4.32	14.78	2.87	21.97
branch, 45 degrees	nr	0.45	4.32	14.78	2.87	21.97
branch, access, 87.5 degrees	nr	0.45	4.32	29.17	5.02	38.51
branch, access, 67.5 degrees	nr	0.45	4.32	29.17	5.02	38.51
branch, access, 45 degrees	nr	0.45	4.32	29.17	5.02	38.51
branch, double, 87.5 degrees	nr	0.45	4.32	29.17	5.02	38.51
branch, double access, 87.5 degrees	nr	0.50	4.80	39.25	6.61	50.66
branch, parallel	nr	0.50	4.80	15.17	3.00	22.97
change piece, to soil	nr	0.40	3.84	8.33	1.83	14.00
offset, 75mm projection	nr	0.40	3.84	9.73	2.04	15.61
offset, 115mm projection	nr	0.40	3.84	12.11	2.39	18.34
offset, 150mm projection	nr	0.40	3.84	12.11	2.39	18.34
offset, 225mm projection	nr	0.40	3.84	15.67	2.93	22.44
offset, 300mm projection	nr	0.40	3.84	17.85	3.25	24.94
roof vent connector for asphalt	nr	0.30	2.88	25.33	4.23	32.44
roof vent connector for felt	nr	0.30	2.88	61.15	9.60	73.63
roof outlet	nr	0.30	2.88	33.05	5.39	41.32
100mm diameter pipe	m	0.55	5.28	16.59	3.28	25.15
Extra over for						
bend, short radius	nr	0.45	4.32	13.59	2.69	20.60

	Unit	Labour hours	Net labour (£)	Net material (£)	O'heads /profit (£)	Total (£)
bend, short radius, access	nr	0.45	4.32	28.75	4.96	38.03
bend, long radius	nr	0.45	4.32	22.03	3.95	30.30
bend, long radius, access	nr	0.45	4.32	37.18	6.22	47.73
bend, long radius, rest	nr	0.45	4.32	26.69	4.65	35.66
branch, 87.5 degrees	nr	0.50	4.80	21.04	3.88	29.72
branch, 76 degrees	nr	0.50	4.80	21.04	3.88	29.72
branch, 67.5 degrees	nr	0.50	4.80	21.04	3.88	29.72
branch, 45 degrees	nr	0.50	4.80	21.04	3.88	29.72
branch, access, 87.5 degrees	nr	0.50	4.80	36.19	6.15	47.14
branch, access, 76 degrees	nr	0.50	4.80	36.19	6.15	47.14
branch, access, 67.5 degrees	nr	0.50	4.80	36.19	6.15	47.14
branch, access, 45 degrees	nr	0.50	4.80	36.19	6.15	47.14
branch, double, 87.5 degrees	nr	0.50	4.80	26.01	4.62	35.43
branch, double, 76 degrees	nr	0.50	4.80	26.01	4.62	35.43
branch, double, 67.5 degrees	nr	0.50	4.80	26.01	4.62	35.43
branch, double, 45 degrees	nr	0.50	4.80	26.01	4.62	35.43
branch, double access, 87.5 degrees	nr	0.50	4.80	41.18	6.90	52.88
branch, double access, 76 degrees	nr	0.50	4.80	41.18	6.90	52.88
branch, double access, 67.5 degrees	nr	0.50	4.80	41.18	6.90	52.88
branch, double access, 45 degrees	nr	0.50	4.80	41.18	6.90	52.88
branch, parallel	nr	0.50	4.80	22.03	4.02	30.85
change piece, to soil	nr	0.40	3.84	10.26	2.11	16.22
change piece, to drain	nr	0.40	3.84	26.69	4.58	35.11
change piece, to clay	nr	0.40	3.84	18.85	3.40	26.09
WC connector	nr	0.45	4.32	14.80	2.87	21.99
WC connecting bend, 300mm tail	nr	0.45	4.32	23.76	4.21	32.29
long tail bend, 87.5 degrees	nr	0.45	4.32	41.18	6.83	52.33
long tail bend, 76 degrees	nr	0.45	4.32	50.65	8.25	63.22
offset, 75mm projection	nr	0.45	4.32	14.29	2.79	21.40
offset, 115mm projection	nr	0.45	4.32	17.21	3.23	24.76
offset, 150mm projection	nr	0.45	4.32	17.06	3.21	24.59
offset, 225mm projection	nr	0.45	4.32	17.08	3.21	24.61
offset, 300mm projection	nr	0.45	4.32	22.03	3.95	30.30
roof vent connector for asphalt	nr	0.35	3.36	29.97	5.00	38.33
roof vent connector for felt	nr	0.35	3.36	74.41	11.67	89.44
roof outlet	nr	0.35	3.36	42.41	6.87	52.64

Soil pipes (cont'd)	Unit	Labour hours	Net labour (£)	Net material (£)	O'heads /profit (£)	Total (£)
150mm diameter pipe	m	0.80	7.68	37.26	6.74	51.68
Extra over for						
bend, short radius	nr	0.70	6.72	24.29	4.65	35.66
bend, short radius, access	nr	0.70	6.72	40.86	7.14	54.72
bend, long radius	nr	0.70	6.72	39.88	6.99	53.59
bend, long radius, access	nr	0.70	6.72	56.42	9.47	72.61
bend, long radius, rest	nr	0.70	6.72	45.64	7.85	60.21
branch, 87.5 degrees	nr	0.70	6.72	43.30	7.50	57.52
branch, 45 degrees	nr	0.80	7.68	43.30	7.65	58.63
branch, access, 87.5 degrees	nr	0.80	7.68	59.84	10.13	77.65
branch, access, 76 degrees	nr	0.80	7.68	56.44	9.62	73.74
branch, access, 45 degrees	nr	0.80	7.68	57.69	9.81	75.18
branch, double, 87.5 degrees	nr	0.80	7.68	66.43	11.12	85.23
branch, double, 76 degrees	nr	0.80	7.68	64.26	10.79	82.73
branch, double access, 87.5 degrees	nr	0.80	7.68	82.98	13.60	104.26
branch, double access, 76 degrees	nr	0.80	7.68	80.82	13.28	101.78
change piece, to soil	nr	0.50	4.80	20.44	3.79	29.03
change piece, to drain	nr	0.50	4.80	32.63	5.61	43.04
change piece, to clay	nr	0.50	4.80	21.21	3.90	29.91
offset, 75mm projection	nr	0.70	6.72	30.16	5.53	42.41
offset, 115mm projection	nr	0.70	6.72	35.34	6.31	48.37
offset, 150mm projection	nr	0.70	6.72	35.34	6.31	48.37
offset, 225mm projection	nr	0.70	6.72	40.16	7.03	53.91
offset, 300mm projection	nr	0.70	6.72	51.58	8.74	67.05
roof vent connector for asphalt	nr	0.50	4.80	55.67	9.07	69.54
roof vent connector for felt	nr	0.50	4.80	111.32	17.42	133.54

	Unit	Labour hours	Net labour (£)	Net material (£)	O'heads /profit (£)	Total (£)
OVERFLOWS						
uPVC, Terrain overflow system, solvent-welded joints, to brickwork with clips at 500mm maximum centres, plugging						
19mm diameter pipe	m	0.19	1.82	1.09	0.44	3.35
Extra over for						
bend, 91 degrees	nr	0.17	1.63	0.88	0.38	2.89
bend, 135 degrees	nr	0.17	1.63	1.03	0.40	3.06
branch, 91 degrees	nr	0.20	1.92	1.03	0.44	3.39
connector, bent tank	nr	0.20	1.92	1.40	0.50	3.82
connector, straight tank	nr	0.20	1.92	1.25	0.48	3.65
connector, iron to uPVC	nr	0.20	1.92	1.16	0.46	3.54
uPVC, Marley overflow system, solvent-welded connector joints, to brickwork with clips at 500mm maximum centres, plugging						
22mm diameter pipe	m	0.19	1.82	0.96	0.42	3.20
Extra over for						
bend, 88.5 degrees	nr	0.17	1.63	0.85	0.37	2.85
bend, adjustable	nr	0.17	1.63	1.15	0.42	3.20
branch, 88.5 degrees	nr	0.20	1.92	0.78	0.41	3.11
adaptor, iron to uPVC	nr	0.17	1.63	0.72	0.35	2.70
adaptor, straight copper to uPVC	nr	0.17	1.63	0.57	0.33	2.53
straight tank connector	nr	0.20	1.92	1.22	0.47	3.61
bent tank connector	nr	0.20	1.92	1.29	0.48	3.69

	Unit	Labour hours	Net labour (£)	Net material (£)	O'heads /profit (£)	Total (£)
ABS, OsmaWeld system, solvent-welded connector joints, to brickwork with clips at 500mm maximum centres, plugging						
19mm diameter pipe	m	0.19	1.82	0.59	0.36	2.77
Extra over for						
bend, 90 degrees	nr	0.17	1.63	0.59	0.33	2.55
tee, 90 degrees	nr	0.20	1.92	0.66	0.39	2.97
connector, bent tank	nr	0.20	1.92	0.88	0.42	3.22
Polypropylene, Osma/ClearBore, ring-sealed socket joints, to brickwork with clips at 500mm maximum centres, plugging						
19mm diameter pipe	m	0.19	1.82	0.52	0.35	2.69
Extra over for						
bend, 90 degrees	nr	0.17	1.63	0.59	0.33	2.55
tee, 90 degrees	nr	0.20	1.92	0.66	0.39	2.97
connector, bent tank	nr	0.20	1.92	0.85	0.42	3.19
connector, straight tank	nr	0.20	1.92	0.85	0.42	3.19
Polypropylene, Hepworth Push-Fit overflow system, ring-sealed socket with clips at 500mm maximum centres, plugging						
19mm diameter pipe	m	0.19	1.82	0.37	0.33	2.52
Extra over for						
bend, 90 degrees	nr	0.17	1.63	0.44	0.31	2.38
tee, 90 degrees	nr	0.20	1.92	0.44	0.35	2.71
reducer	nr	0.17	1.63	0.44	0.31	2.38
connector, straight tank	nr	0.20	1.92	0.44	0.35	2.71
connector, bent tank	nr	0.20	1.92	0.44	0.35	2.71

Overflows (cont'd)	Unit	Labour hours	Net labour (£)	Net material (£)	O'heads /profit (£)	Total (£)
ABS, Hepworth overflow system, ring solvent-welded connector joints, to brickwork with clips at 500mm maximum centres, plugging						
19mm diameter pipe	m	0.19	1.82	0.37	0.33	2.52
Extra over for						
bend, 90 degrees	nr	0.17	1.63	0.44	0.31	2.38
tee, 90 degrees	nr	0.20	1.92	0.44	0.35	2.71
reducer	nr	0.17	1.63	0.44	0.31	2.38
connector, straight tank	nr	0.20	1.92	0.44	0.35	2.71
connector, bent tank	nr	0.20	1.92	0.44	0.35	2.71
TRAPS						
Polypropylene, Terrain, screwed joint to outlet and pipe						
bottle P trap						
32mm	nr	0.27	2.59	4.17	1.01	7.77
40mm	nr	0.32	3.07	4.99	1.21	9.27
bottle S trap						
32mm	nr	0.27	2.59	5.03	1.14	8.76
40mm	nr	0.32	3.07	6.12	1.38	10.57
self-resealing bottle P trap						
32mm	nr	0.27	2.59	5.20	1.17	8.96
40mm	nr	0.32	3.07	6.09	1.37	10.53
self-resealing bottle S trap						
32mm	nr	0.27	2.59	6.14	1.31	10.04
40mm	nr	0.32	3.07	7.11	1.53	11.71
telescopic bottle P trap						
32mm	nr	0.27	2.59	7.10	1.45	11.14
40mm	nr	0.32	3.07	8.54	1.74	13.35

Traps (cont'd)	Unit	Labour hours	Net labour (£)	Net material (£)	O'heads /profit (£)	Total (£)
tubular P trap						
32mm	nr	0.27	2.59	3.75	0.95	7.29
40mm	nr	0.32	3.07	4.33	1.11	8.51
50mm	nr	0.35	3.36	10.37	2.06	15.79
tubular S trap						
32mm	nr	0.27	2.59	4.76	1.10	8.45
40mm	nr	0.32	3.07	5.57	1.30	9.94
washing machine trap						
40mm	nr	0.40	3.84	7.52	1.70	13.06
tubular running P trap						
32mm	nr	0.30	2.88	5.76	1.30	9.94
40mm	nr	0.32	3.07	6.28	1.40	10.75
50mm	nr	0.35	3.36	14.67	2.70	20.73
tubular P trap for bath						
40mm	nr	0.33	3.17	8.05	1.68	12.90
tubular P trap assembly for bath						
40mm, white	nr	0.50	4.80	13.72	2.78	21.30
40mm, chromium-plated	nr	0.50	4.80	16.44	3.19	24.43
tubular P lowline bath trap						
40mm	nr	0.50	4.80	4.70	1.43	10.93
tubular P lowline bath trap with overflow						
40mm	nr	0.50	4.80	7.26	1.81	13.87

	Unit	Labour hours	Net labour (£)	Net material (£)	O'heads /profit (£)	Total (£)
Polypropylene, Marley, screwed joint to outlet and pipe						
tubular P trap						
32mm	nr	0.27	2.59	2.70	0.79	6.08
40mm	nr	0.32	3.07	3.24	0.95	7.26
tubular S trap						
32mm	nr	0.27	2.59	3.24	0.87	6.70
40mm	nr	0.32	3.07	3.85	1.04	7.96
standard bottle trap						
32mm	nr	0.25	2.40	2.77	0.78	5.95
40mm	nr	0.30	2.88	2.59	0.82	6.29
monitor anti-syphon bottle trap						
32mm	nr	0.25	2.40	4.39	1.02	7.81
40mm	nr	0.30	2.88	4.58	1.12	8.58
low inlet tubular P trap, 40mm						
plain	nr	0.33	3.17	3.24	0.96	7.37
with access	nr	0.33	3.17	3.78	1.04	7.99
with overflow and outlet	nr	0.50	4.80	9.34	2.12	16.26
shallow bath trap, 40mm	nr	0.33	3.17	4.08	1.09	8.34
low level bath P trap	nr	0.35	3.36	3.81	1.08	8.25
washing machine kit	nr	0.40	3.84	5.46	1.40	10.70
running P trap						
40mm	nr	0.33	3.17	4.64	1.17	8.98
50mm	nr	0.33	3.17	10.98	2.12	16.27

Traps (cont'd)	Unit	Labour hours	Net labour (£)	Net material (£)	O'heads /profit (£)	Total (£)
Polypropylene, Osma, screwed joint to outlet and pipe						
bottle trap with 38mm seal						
32mm	nr	0.25	2.40	2.28	0.70	5.38
40mm	nr	0.30	2.88	2.84	0.86	6.58
bottle trap with 76mm seal						
32mm	nr	0.25	2.40	2.62	0.75	5.77
40mm	nr	0.30	2.88	3.12	0.90	6.90
tubular P trap						
32mm	nr	0.27	2.59	2.43	0.75	5.77
40mm	nr	0.32	3.07	2.92	0.90	6.89
tubular S trap						
32mm	nr	0.27	2.59	2.91	0.82	6.33
40mm	nr	0.32	3.07	3.54	0.99	7.60
adjustable tubular P trap						
32mm	nr	0.30	2.88	2.96	0.88	6.72
40mm	nr	0.35	3.36	3.49	1.03	7.88
running P trap						
32mm	nr	0.30	2.88	3.66	0.98	7.52
40mm	nr	0.32	3.07	4.00	1.06	8.13
bath trap, shallow seal						
32mm	nr	0.30	2.88	2.16	0.76	5.80
40mm	nr	0.32	3.07	2.34	0.81	6.22
bath trap, 76mm seal						
40mm	nr	0.33	3.17	3.28	0.97	7.42

	Unit	Labour hours	Net labour (£)	Net material (£)	O'heads /profit (£)	Total (£)
bath trap and overflow						
40mm	nr	0.50	4.80	7.41	1.83	14.04
washing machine half trap						
40mm	nr	0.40	3.84	5.33	1.38	10.55
washing machine trap						
40mm	nr	0.45	4.32	4.60	1.34	10.26
Polypropylene, Hepworth, screwed joint to outlet and pipe						
bottle trap with 38mm seal						
32mm	nr	0.25	2.40	1.95	0.65	5.00
40mm	nr	0.30	2.88	2.24	0.77	5.89
bottle trap with 76mm seal						
32mm	nr	0.25	2.40	1.87	0.64	4.91
40mm	nr	0.30	2.88	2.46	0.80	6.14
tubular P trap						
32mm	nr	0.27	2.59	1.72	0.65	4.96
40mm	nr	0.32	3.07	2.01	0.76	5.84
tubular S trap						
32mm	nr	0.27	2.59	2.43	0.75	5.77
40mm	nr	0.32	3.07	2.84	0.89	6.80
adjustable tubular P trap						
32mm	nr	0.30	2.88	2.15	0.75	5.78
40mm	nr	0.35	3.36	2.28	0.85	6.49
running P trap						
40mm	nr	0.32	3.07	3.42	0.97	7.46

	Unit	Material supply (£)	Material waste (%)	Total (£)
R12 DRAINAGE BELOW GROUND				
MATERIAL				
Cast iron to BS437, Drainage Systems, spigot and socket, caulked lead joints				
100mm diameter pipe	m	26.00	5.0	27.30
bend, short radius	nr	21.83	2.5	22.38
bend, short radius with access door	nr	35.18	2.5	36.06
bend, long radius	nr	37.33	2.5	38.26
bend, short radius with duckfoot	nr	27.68	2.5	28.37
bend, long radius with duckfoot	nr	40.92	2.5	41.94
branch, 100 x 100mm	nr	34.55	2.5	35.41
branch, double	nr	127.27	2.5	130.45
connector, drain to soil	nr	24.34	2.5	24.95
150mm diameter pipe	m	44.68	5.0	46.91
bend, short radius	nr	64.14	2.5	65.74
bend, short radius with access door	nr	80.08	2.5	82.08
bend, long radius	nr	69.29	2.5	71.02
bend, short radius with duckfoot	nr	61.98	2.5	63.53
branch, 150 x 150mm	nr	68.78	2.5	70.50
branch, double	nr	163.43	2.5	167.52
225mm diameter pipe	m	205.69	5.0	215.97
bend, short radius	nr	180.30	2.5	184.81
bend, short radius with access door	nr	276.68	2.5	283.60
bend, long radius	nr	330.55	2.5	338.81
bend, short radius with duckfoot	nr	216.94	2.5	222.36
branch, 225 to 150mm	nr	219.23	2.5	224.71
branch, 225 to 225mm	nr	234.57	2.5	240.43
connector, drain to soil	nr	106.45	2.5	109.11

	Unit	Material supply (£)	Material waste (%)	Total (£)
Cast iron to BS437, Drainage Castings Timesaver system, flexible joints				
50mm diameter pipe	m	20.61	5.0	21.64
coupling	nr	10.22	2.5	10.48
stepped coupling	nr	9.75	2.5	9.99
bend, 87.5 degrees	nr	18.34	2.5	18.80
bend, 45 degrees	nr	17.28	2.5	17.71
branch, 50 x 50mm	nr	25.49	2.5	26.13
75mm diameter pipe	m	20.02	5.0	21.02
coupling	nr	10.22	2.5	10.48
stepped coupling	nr	9.75	2.5	9.99
bend, 87.5 degrees	nr	18.34	2.5	18.80
bend, 45 degrees	nr	17.28	2.5	17.71
branch, 75 x 75mm	nr	25.49	2.5	26.13
diminishing pipe, 75 to 50mm	nr	10.99	2.5	11.26
gulley trap, 87.5 degrees	nr	19.22	2.5	19.70
gulley trap, vertical	nr	30.43	2.5	31.19
100mm diameter pipe	m	19.35	5.0	20.32
coupling	nr	12.18	2.5	12.48
stepped coupling	nr	11.43	2.5	11.72
bend, 87.5 degrees	nr	22.58	2.5	23.14
bend, 80 degrees	nr	22.58	2.5	23.14
bend, 67.5 degrees	nr	20.90	2.5	21.42
bend, 60 degrees	nr	20.90	2.5	21.42
bend, 45 degrees	nr	19.22	2.5	19.70
bend, 35 degrees	nr	19.22	2.5	19.70
bend, 22.5 degrees	nr	17.55	2.5	17.99
bend, 11 degrees	nr	11.17	2.5	11.45
bend, 10 degrees	nr	15.88	2.5	16.28
bend, 87.5 degrees, heel rest	nr	25.93	2.5	26.58
bend, access rear	nr	48.67	2.5	49.89
bend, access side	nr	59.36	2.5	60.84
bend, long radius	nr	27.69	2.5	28.38
bend, long radius, heel rest	nr	40.14	2.5	41.14
bend, long tail	nr	53.62	2.5	54.96
bend, large radius	nr	104.61	2.5	107.23

Cast iron drains (cont'd)	Unit	Material supply (£)	Material waste (%)	Total (£)
bend, WC socket, heel rest	nr	35.20	2.5	36.08
branch, 100 x 100mm	nr	31.17	2.5	31.95
branch, access	nr	69.16	2.5	70.89
branch, long tail	nr	59.87	2.5	61.37
branch, double, 100 x 100mm	nr	45.17	2.5	46.30
branch, double with access, 100 x 100mm	nr	83.44	2.5	85.53
diminishing pipe, 100 x 50mm	nr	18.16	2.5	18.61
diminishing pipe, 100 x 75mm	nr	19.22	2.5	19.70
transitional pipe, socket for clayware	nr	19.78	2.5	20.27
transitional pipe, socket for cast iron	nr	27.06	2.5	27.74
gulley trap, 87.5 degrees	nr	30.00	2.5	30.75
gulley trap, 45 degrees	nr	34.91	2.5	35.78
gulley trap, vertical	nr	34.91	2.5	35.78
gulley trap, surface access, 87.5 degrees	nr	59.80	2.5	61.29
gulley trap, surface access, vertical	nr	67.59	2.5	69.28
running trap, plain	nr	71.67	2.5	73.46
running trap, single access	nr	107.90	2.5	110.60
running trap, double access	nr	149.41	2.5	153.15
gulley inlet, plain	nr	31.21	2.5	31.99
gulley inlet, sealed	nr	45.59	2.5	46.73
gulley inlet, horizontal branch plain	nr	34.02	2.5	34.87
gulley inlet, horizontal branch sealed	nr	48.40	2.5	49.61
gulley inlet, vertical branch plain	nr	39.16	2.5	40.14
gulley inlet, vertical branch sealed	nr	54.05	2.5	55.40
raising piece, 75mm	nr	18.79	2.5	19.26
raising piece, 115mm	nr	20.83	2.5	21.35
raising piece, 150mm	nr	22.48	2.5	23.04
raising piece, 225mm	nr	28.59	2.5	29.30
raising piece, 305mm	nr	33.36	2.5	34.19
raising piece, 380mm	nr	46.38	2.5	47.54
raising piece, 455mm	nr	52.74	2.5	54.06
gulley trap, 425mm deep, 87.5 degrees outlet	nr	66.94	2.5	68.61
garage gulley, 325 x 600mm	nr	351.10	2.5	359.88

	Unit	Material supply (£)	Material waste (%)	Total (£)
150mm diameter pipe	m	36.46	5.0	38.28
coupling	nr	15.15	2.5	15.53
stepped coupling	nr	13.26	2.5	13.59
bend, 87.5 degrees	nr	45.59	2.5	46.73
bend, 67.5 degrees	nr	42.33	2.5	43.39
bend, 45 degrees	nr	38.80	2.5	39.77
bend, 35 degrees	nr	38.80	2.5	39.77
bend, 22.5 degrees	nr	35.70	2.5	36.59
bend, 10 degrees	nr	32.38	2.5	33.19
bend, 87.5 degrees, heel rest	nr	51.23	2.5	52.51
bend, access rear	nr	110.24	2.5	113.00
bend, access side	nr	110.24	2.5	113.00
bend, long radius	nr	59.36	2.5	60.84
bend, long radius, heel rest	nr	60.24	2.5	61.75
branch, 150 x 150mm	nr	64.90	2.5	66.52
branch, 150 x 100mm	nr	64.74	2.5	66.36
branch access, 150 x 150mm	nr	127.87	2.5	131.07
branch access, 150 x 100m	nr	127.67	2.5	130.86
diminishing pipe, 150 x 100mm	nr	29.46	2.5	30.20
transitional pipe, socket for clayware	nr	32.07	2.5	32.87
transitional pipe, socket for cast iron	nr	34.11	2.5	34.96
gulley trap, 87.5 degrees	nr	74.62	2.5	76.49
gulley trap, 45 degrees	nr	102.30	2.5	104.86
gulley trap, vertical	nr	102.30	2.5	104.86
225mm diameter pipe	m	92.11	5.0	96.72
coupling	nr	33.53	2.5	34.37
bend, 87.5 degrees	nr	157.77	2.5	161.71
bend, 45 degrees	nr	38.80	2.5	39.77
bend, 22.5 degrees	nr	86.02	2.5	88.17
bend, 10 degrees	nr	157.77	2.5	161.71
bend, access rear	nr	241.70	2.5	247.74
bend, access side	nr	235.80	2.5	241.70
branch, 225 x 225mm	nr	227.16	2.5	232.84
branch, 225 x 100mm	nr	191.63	2.5	196.42
branch, 225 x 150mm	nr	204.34	2.5	209.45
branch, access, 225 x 100mm	nr	288.85	2.5	296.07
branch, access, 225 x 150mm	nr	298.04	2.5	305.49
branch, double, 225 x 100mm	nr	298.04	2.5	305.49
branch, double, 225 x 150mm	nr	302.45	2.5	310.01

Cast iron drains (cont'd)	Unit	Material supply (£)	Material waste (%)	Total (£)
branch, double, 225 x 225mm	nr	328.84	2.5	337.06
branch, double, with access 225 x 100mm	nr	378.51	2.5	387.97
branch, double, with access 225 x 150mm	nr	382.78	2.5	392.35
branch, double, with access 225 x 225mm	nr	408.29	2.5	418.50
diminishing pipe, 225 x 100mm	nr	75.52	2.5	77.41
diminishing pipe, 225 x 150mm	nr	75.52	2.5	77.41
diminishing pipe, 225 x 200mm	nr	58.64	2.5	60.11
transitional pipe, pipe for clayware	nr	109.67	2.5	112.41

	Unit	Labour hours	Net labour (£)	Net material (£)	O'heads /profit (£)	Total (£)
R12 DRAINAGE BELOW GROUND						
Cast iron to BS437, Drainage Systems, spigot and socket, caulked lead joints, in trenches prepared by others						
100mm diameter pipe						
laid straight	m	0.65	6.24	28.61	5.23	40.08
laid vertically	m	1.00	9.60	31.47	6.16	47.23
in lengths less than 3m	m	0.95	9.12	32.90	6.30	48.32
Extra over for						
bend, short radius	nr	0.55	5.28	21.83	4.07	31.18
bend, short radius with access door	nr	0.55	5.28	35.18	6.07	46.53
bend, long radius	nr	0.55	5.28	37.33	6.39	49.00
bend, short radius with duckfoot	nr	0.55	5.28	27.68	4.94	37.90
bend, long radius with duckfoot	nr	0.55	5.28	40.92	6.93	53.13
branch, 100 x 100mm	nr	0.80	7.68	34.55	6.33	48.56
branch, double	nr	1.05	10.08	127.27	20.60	157.95
connector, drain to soil	nr	0.60	5.76	24.34	4.51	34.61
150mm diameter pipe						
laid straight	m	0.90	8.64	48.75	8.61	66.00
laid vertical	m	0.70	6.72	53.63	9.05	69.40
in lengths less than 3m	m	1.26	12.10	56.06	10.22	78.38
Extra over for						
bend, short radius	nr	0.70	6.72	64.14	10.63	81.49
bend, short radius with access door	nr	0.70	6.72	80.08	13.02	99.82
bend, long radius	nr	0.70	6.72	69.29	11.40	87.41
bend, short radius with duckfoot	nr	0.70	6.72	61.98	10.30	79.00
branch, 150 x 150mm	nr	1.05	10.08	68.67	11.81	90.56
branch, double	nr	1.05	10.08	163.43	26.03	199.54

Cast iron drains (cont'd)	Unit	Labour hours	Net labour (£)	Net material (£)	O'heads /profit (£)	Total (£)
225mm diameter pipe						
laid straight	m	1.30	12.48	218.34	34.62	265.44
laid vertical	m	1.50	14.40	240.17	38.19	292.76
in lengths less than 3m	m	1.70	16.32	251.09	40.11	307.52
Extra over for						
bend, short radius	nr	1.25	12.00	180.30	28.84	221.14
bend, short radius with access door	nr	1.25	12.00	276.68	43.30	331.98
bend, long radius	nr	1.25	12.00	330.55	51.38	393.93
bend, short radius with duckfoot	nr	1.25	12.00	216.94	34.34	263.28
branch, 225 to 150mm	nr	1.70	16.32	219.23	35.33	270.88
branch, 225 to 225mm	nr	1.70	16.32	234.57	37.63	288.52
connector, drain to soil	nr	0.95	9.12	106.45	17.34	132.91
Cast iron to BS437, Drainage Castings Timesaver system, flexible joints, in trenches prepared by others						
50mm diameter pipe						
laid straight	m	0.30	2.88	24.25	4.07	31.20
laid vertically	m	0.65	6.24	26.68	4.94	37.86
lengths less than 3m	m	0.60	5.76	27.84	5.04	38.64
Extra over for						
coupling	nr	0.20	1.92	10.22	1.82	13.96
stepped coupling	nr	0.20	1.92	9.75	1.75	13.42
bend, 87.5 degrees	nr	0.30	2.88	18.34	3.18	24.40
bend, 45 degrees	nr	0.30	2.88	17.28	3.02	23.18
branch, 50 x 50mm	nr	0.30	2.88	25.49	4.26	32.63
75mm diameter pipe						
laid straight	m	0.35	3.36	22.76	3.92	30.04
laid vertically	m	0.70	6.72	25.04	4.76	36.52
in lengths less than 3m	m	0.65	6.24	26.17	4.86	37.27

	Unit	Labour hours	Net labour (£)	Net material (£)	O'heads /profit (£)	Total (£)
Extra over for						
coupling	nr	0.25	2.40	10.22	1.89	14.51
stepped coupling	nr	0.25	2.40	9.75	1.82	13.97
bend, 87.5 degrees	nr	0.35	3.36	18.34	3.25	24.95
bend, 45 degrees	nr	0.35	3.36	17.28	3.10	23.74
branch, 75 x 75mm	nr	0.60	5.76	25.49	4.69	35.94
diminishing pipe, 75 to 50mm	nr	0.30	2.88	10.99	2.08	15.95
gulley trap, 87.5 degrees	nr	0.70	6.72	19.22	3.89	29.83
gulley trap, vertical	nr	0.70	6.72	30.43	5.57	42.72
100mm diameter pipe						
laid straight	m	0.40	3.84	22.40	3.94	30.18
laid vertically	m	0.75	7.20	24.64	4.78	36.62
in lengths less than 3m	m	0.70	6.72	25.76	4.87	37.35
Extra over for						
coupling	nr	0.40	3.84	12.18	2.40	18.42
stepped coupling	nr	0.40	3.84	11.43	2.29	17.56
bend, 87.5 degrees	nr	0.40	3.84	22.58	3.96	30.38
bend, 80 degrees	nr	0.40	3.84	22.58	3.96	30.38
bend, 67.5 degrees	nr	0.40	3.84	20.90	3.71	28.45
bend, 60 degrees	nr	0.40	3.84	20.90	3.71	28.45
bend, 45 degrees	nr	0.40	3.84	19.22	3.46	26.52
bend, 35 degrees	nr	0.40	3.84	19.22	3.46	26.52
bend, 22.5 degrees	nr	0.40	3.84	17.55	3.21	24.60
bend, 11 degrees	nr	0.40	3.84	19.37	3.48	26.69
bend, 10 degrees	nr	0.40	3.84	15.88	2.96	22.68
bend, 87.5 degrees, heel rest	nr	0.40	3.84	25.93	4.47	34.24
bend, access rear	nr	0.40	3.84	48.67	7.88	60.39
bend, access side	nr	0.40	3.84	59.36	9.48	72.68
bend, long radius	nr	0.45	4.32	27.69	4.80	36.81
bend, long radius, heel rest	nr	0.45	4.32	40.14	6.67	51.13
bend, long tail	nr	0.45	4.32	53.62	8.69	66.63
bend, large radius	nr	0.45	4.32	104.58	16.34	125.24
bend, WC socket, heel rest	nr	0.45	4.32	35.20	5.93	45.45
branch, 100 x 100mm	nr	0.60	5.76	31.17	5.54	42.47
branch, access	nr	0.60	5.76	69.16	11.24	86.16
branch, long tail	nr	0.65	6.24	59.87	9.92	76.03
branch, double, 100 x 100mm	nr	0.75	7.20	45.17	7.86	60.23
branch, double with access, 100 x 100mm	nr	0.75	7.20	83.44	13.60	104.24

Cast iron drains (cont'd)	Unit	Labour hours	Net labour (£)	Net material (£)	O'heads /profit (£)	Total (£)
diminishing pipe, 100 x 50mm	nr	0.40	3.84	18.16	3.30	25.30
diminishing pipe, 100 x 75mm	nr	0.40	3.84	19.22	3.46	26.52
transitional pipe, socket for clayware	nr	0.40	3.84	19.78	3.54	27.16
transitional pipe, socket for cast iron	nr	0.40	3.84	27.06	4.63	35.53
gulley trap, 87.5 degrees	nr	0.70	6.72	30.00	5.51	42.23
gulley trap, 45 degrees	nr	0.70	6.72	34.91	6.24	47.87
gulley trap, vertical	nr	0.70	6.72	34.91	6.24	47.87
gulley trap, surface access, 87.5 degrees	nr	0.70	6.72	59.80	9.98	76.50
gulley trap, surface access, vertical	nr	0.70	6.72	67.59	11.15	85.46
running trap, plain	nr	0.70	6.72	71.67	11.76	90.15
running trap, single access	nr	0.70	6.72	107.90	17.19	131.81
running trap, double access	nr	0.70	6.72	149.41	23.42	179.55
gulley inlet, plain	nr	0.90	8.64	31.21	5.98	45.83
gulley inlet, sealed	nr	0.90	8.64	45.59	8.13	62.36
gulley inlet, horizontal branch plain	nr	0.90	8.64	34.02	6.40	49.06
gulley inlet, horizontal branch sealed	nr	0.90	8.64	48.40	8.56	65.60
gulley inlet, vertical branch plain	nr	0.90	8.64	39.16	7.17	54.97
gulley inlet, vertical branch sealed	nr	0.91	8.74	53.54	9.34	71.62
raising piece, 75mm	nr	0.40	3.84	18.79	3.39	26.02
raising piece, 115mm	nr	0.40	3.84	20.83	3.70	28.37
raising piece, 150mm	nr	0.40	3.84	22.48	3.95	30.27
raising piece, 225mm	nr	0.40	3.84	28.59	4.86	37.29
raising piece, 305mm	nr	0.40	3.84	33.36	5.58	42.78
raising piece, 380mm	nr	0.50	4.80	46.38	7.68	58.86
raising piece, 455mm	nr	0.50	4.80	52.74	8.63	66.17
gulley trap, 425mm deep, 87.5 degrees outlet	nr	1.50	14.40	107.94	18.35	140.69
garage gulley, 325 x 600mm	nr	2.40	23.04	351.10	56.12	430.26
150mm diameter pipe						
laid straight	m	0.60	5.76	40.78	6.98	53.52
laid vertically	m	0.80	7.68	44.86	7.88	60.42
in lengths less than 3m	m	0.90	8.64	46.90	8.33	63.87

	Unit	Labour hours	Net labour (£)	Net material (£)	O'heads /profit (£)	Total (£)
Extra over for						
coupling	nr	0.60	5.76	14.64	3.06	23.46
stepped coupling	nr	0.60	5.76	13.26	2.85	21.87
bend, 87.5 degrees	nr	0.60	5.76	45.59	7.70	59.05
bend, 67.5 degrees	nr	0.60	5.76	42.33	7.21	55.30
bend, 45 degrees	nr	0.60	5.76	38.80	6.68	51.24
bend, 35 degrees	nr	0.60	5.76	38.80	6.68	51.24
bend, 22.5 degrees	nr	0.60	5.76	35.70	6.22	47.68
bend, 10 degrees	nr	0.60	5.76	32.38	5.72	43.86
bend, 87.5 degrees, heel rest	nr	0.60	5.76	51.23	8.55	65.54
bend, access rear	nr	0.60	5.76	110.24	17.40	133.40
bend, access side	nr	0.60	5.76	110.24	17.40	133.40
bend, long radius	nr	0.65	6.24	59.36	9.84	75.44
bend, long radius, heel rest	nr	0.65	6.24	60.24	9.97	76.45
branch, 150 x 150mm	nr	0.75	7.20	64.90	10.82	82.92
branch, 150 x 100mm	nr	0.75	7.20	64.74	10.79	82.73
branch access, 150 x 150mm	nr	0.75	7.20	127.87	20.26	155.33
branch access, 150 x 100m	nr	0.75	7.20	127.67	20.23	155.10
diminishing pipe, 150 x 100mm	nr	0.60	5.76	29.46	5.28	40.50
transitional pipe, socket for clayware	nr	0.60	5.76	32.07	5.67	43.50
transitional pipe, socket for cast iron	nr	0.60	5.76	34.11	5.98	45.85
gulley trap, 87.5 degrees	nr	1.00	9.60	74.62	12.63	96.85
gulley trap, 45 degrees	nr	1.00	9.60	102.30	16.79	128.69
gulley trap, vertical	nr	1.00	9.60	102.30	16.79	128.69
225mm diameter pipe						
laid straight	m	1.10	10.56	99.96	16.58	127.10
laid vertically	m	1.30	12.48	109.96	18.37	140.81
in lengths less than 3m	m	1.50	14.40	114.95	19.40	148.75
Extra over for						
coupling	nr	1.10	10.56	33.53	6.61	50.70
bend, 87.5 degrees	nr	1.00	9.60	157.77	25.11	192.48
bend, 45 degrees	nr	1.00	9.60	157.77	25.11	192.48
bend, 22.5 degrees	nr	1.00	9.60	157.77	25.11	192.48
bend, 10 degrees	nr	1.00	9.60	157.77	25.11	192.48
bend, access rear	nr	1.00	9.60	241.70	37.70	289.00
bend, access side	nr	1.00	9.60	235.80	36.81	282.21
branch, 225 x 225mm	nr	1.45	13.92	227.16	36.16	277.24

Cast iron drains (cont'd)	Unit	Labour hours	Net labour (£)	Net material (£)	O'heads /profit (£)	Total (£)
branch, 225 x 100mm	nr	1.45	13.92	191.63	30.83	236.38
branch, 225 x 150mm	nr	1.45	13.92	204.34	32.74	251.00
branch, access, 225 x 100mm	nr	1.45	13.92	288.85	45.42	348.19
branch, access, 225 x 150mm	nr	1.45	13.92	298.04	46.79	358.75
branch, double, 225 x 100mm	nr	1.90	18.24	298.04	47.44	363.72
branch, double, 225 x 150mm	nr	1.90	18.24	302.45	48.10	368.79
branch, double, 225 x 225mm	nr	1.90	18.24	328.84	52.06	399.14
branch, double, with access 225 x 100mm	nr	1.90	18.24	378.51	59.51	456.26
branch, double, with access 225 x 150mm	nr	1.90	18.24	382.78	60.15	461.17
branch, double, with access 225 x 225mm	nr	1.90	18.24	408.29	63.98	490.51
diminishing pipe, 225 x 100mm	nr	1.10	10.56	75.52	12.91	98.99
diminishing pipe, 225 x 150mm	nr	1.10	10.56	75.52	12.91	98.99
diminishing pipe, 225 x 200mm	nr	1.10	10.56	58.64	10.38	79.58
transitional pipe, pipe for clayware	nr	0.80	7.68	111.72	17.91	137.31

S

PIPED SUPPLY SYSTEMS

	Unit	Material supply (£)	Material waste (%)	Total (£)
S10 HOT AND COLD WATER				
PIPEWORK				
Copper pipe to BS2871, Table W				
8mm diameter	m	0.84	5.0	0.88
10mm diameter	m	0.90	5.0	0.95
Copper pipe to BS2871, Table X				
15mm diameter	m	1.23	5.0	1.29
22mm diameter	m	2.30	5.0	2.42
28mm diameter	m	2.97	5.0	3.12
35mm diameter	m	6.94	5.0	7.29
42mm diameter	m	8.41	5.0	8.83
54mm diameter	m	11.44	5.0	12.01
Copper pipe to BS2871, Table Z				
15mm diameter	m	1.18	5.0	1.24
22mm diameter	m	2.03	5.0	2.13
28mm diameter	m	2.97	5.0	3.12
35mm diameter	m	5.84	5.0	6.13
42mm diameter	m	7.23	5.0	7.59
54mm diameter	m	9.31	5.0	9.78
Copper pipe to BS2871, Table Y				
15mm diameter	m	2.22	5.0	2.33
18mm diameter	m	2.81	5.0	2.95

Pipework (cont'd)	Unit	Material supply (£)	Material waste (%)	Total (£)
22mm diameter	m	3.88	5.0	4.07
28mm diameter	m	5.65	5.0	5.93
35mm diameter	m	8.90	5.0	9.35
42mm diameter	m	10.77	5.0	11.31
54mm diameter	m	18.32	5.0	19.24
Medium-density polyethylene pipes (MDPE) to BS6572 and BS6370				
20mm diameter	m	0.63	5.0	0.66
25mm diameter	m	0.77	5.0	0.81
32mm diameter	m	1.25	5.0	1.31
50mm diameter	m	3.14	5.0	3.30
63mm diameter	m	4.61	5.0	4.84
FITTINGS				
Lead-free pre-soldered capillary joints and fittings to BS864				
8mm fittings				
straight coupling	nr	0.63	2.5	0.65
reduced coupling, 8 x 6mm	nr	1.96	2.5	2.01
straight female connector, 8mm x 1/4in	nr	1.85	2.5	1.90
straight male connector, 8mm x 1/4in	nr	1.31	2.5	1.34
reducer, 8 x 6mm	nr	1.61	2.5	1.65
elbow, 8mm	nr	1.52	2.5	1.56
street elbow, 8mm	nr	2.11	2.5	2.16
equal tee, 8mm	nr	2.89	2.5	2.96
stop end, 8mm	nr	1.08	2.5	1.11

MATERIAL COSTS

	Unit	Material supply (£)	Material waste (%)	Total (£)
10mm fittings				
straight coupling	nr	0.63	2.5	0.65
reduced coupling, 10 x 8mm	nr	1.27	2.5	1.30
straight male connector, 10mm x 3/8in	nr	1.43	2.5	1.47
male reducing connector, 10mm x 1/2in	nr	2.70	2.5	2.77
reducer, 10 x 8mm	nr	1.18	2.5	1.21
elbow, 10mm	nr	1.41	2.5	1.45
street elbow, 10mm	nr	2.11	2.5	2.16
equal tee, 10mm	nr	2.70	2.5	2.77
stop end, 10mm	nr	1.08	2.5	1.11
15mm fittings				
straight coupling	nr	0.19	2.5	0.19
reduced coupling, 15 x 8mm	nr	1.50	2.5	1.54
reduced coupling, 15 x 10mm	nr	1.13	2.5	1.16
reduced coupling, 15 x 12mm	nr	1.21	2.5	1.24
adaptor coupling, imperial to metric, 1/2in x 15mm	nr	0.88	2.5	0.90
straight female connector, 15mm x 1/2in	nr	1.31	2.5	1.34
female reducing connector, 15mm x 3/8in	nr	2.18	2.5	2.23
female reducing connector, 15mm x 3/4in	nr	3.61	2.5	3.70
straight male connector, 15mm x 1/2in	nr	1.06	2.5	1.09
male reducing connector, 15mm x 3/4in	nr	2.57	2.5	2.63
male reducing connector, 15mm x 3/8in	nr	1.78	2.5	1.82
lead connector	nr	0.85	2.5	0.87
tank connector, 15mm x 1/2in	nr	2.70	2.5	2.77
tank connector, 15mm x 3/4in	nr	6.05	2.5	6.20
reducer, 15 x 8mm	nr	0.92	2.5	0.94
reducer, 15 x 10mm	nr	0.92	2.5	0.94
reducer, 15 x 12mm	nr	0.91	2.5	0.93
female adaptor, 15mm x 1/2in	nr	2.05	2.5	2.10
male adaptor, 15mm x 1/2in	nr	2.05	2.5	2.10
adaptor, imperial to metric, 1/2in x 15mm	nr	0.80	2.5	0.82

Capillary fittings (cont'd)	Unit	Material supply (£)	Material waste (%)	Total (£)
street elbow	nr	1.23	2.5	1.26
male elbow, 15mm x 1/2in	nr	2.38	2.5	2.44
reducing male elbow, 15mm x 3/8in	nr	3.42	2.5	3.51
female elbow, 15mm x 1/2in	nr	2.23	2.5	2.29
reducing female elbow, 15mm x 1/4in	nr	3.36	2.5	3.44
backplate elbow, 15mm x 1/2in	nr	2.62	2.5	2.69
flanged bend, 15mm x 1/2in	nr	4.85	2.5	4.97
flanged bend, 15mm x 3/4in	nr	4.85	2.5	4.97
overflow bend, 15mm x 1/2in	nr	5.94	2.5	6.09
slow bend	nr	1.46	2.5	1.50
return bend	nr	5.23	2.5	5.36
obtuse elbow	nr	1.20	2.5	1.23
tee, reduced branch, largest end, 15mm	nr	2.76	2.5	2.83
tee, one end and one branch reduced, largest end, 15mm	nr	3.07	2.5	3.15
tee, both ends reduced, largest end, 15mm	nr	3.83	2.5	3.93
corner tee	nr	5.36	2.5	5.49
sweep tee, 90 degrees	nr	3.54	2.5	3.63
offset tee, 15mm equal	nr	5.59	2.5	5.73
double sweep tee, 15mm	nr	4.87	2.5	4.99
cross equal tee, 15mm	nr	4.49	2.5	4.60
stop end, 15mm	nr	1.60	2.5	1.64
tap connector, 15mm x 3/4in	nr	2.29	2.5	2.35
bent union adaptor, 15mm x 3/4in	nr	2.65	2.5	2.72
bent male union connector, 15mm x 1/2in	nr	4.76	2.5	4.88
bent female union connector, 15mm x 1/2in	nr	4.76	2.5	4.88
22mm fittings				
straight coupling	nr	0.42	2.5	0.43
reduced coupling, 22 x 8mm	nr	1.78	2.5	1.82
reduced coupling, 22 x 10mm	nr	1.78	2.5	1.82
reduced coupling, 22 x 15mm	nr	1.01	2.5	1.04
straight female connector, 22mm x 3/4in	nr	2.31	2.5	2.37
female reducing connector, 22mm x 1/2in	nr	3.44	2.5	3.53

MATERIAL COSTS

	Unit	Material supply (£)	Material waste (%)	Total (£)
female reducing connector, 22mm x 1in	nr	4.76	2.5	4.88
straight male connector, 22mm x 3/4in	nr	4.65	2.5	4.77
male reducing connector, 22mm x 1/2in	nr	3.25	2.5	3.33
male reducing connector, 22mm x 1in	nr	3.21	2.5	3.29
lead connector	nr	1.31	2.5	1.34
tank connector, 22mm x 1in	nr	9.66	2.5	9.90
reducer, 22 x 12mm	nr	1.63	2.5	1.67
female adaptor, 22mm x 3/4in	nr	3.05	2.5	3.13
male adaptor, 22mm x 3/4in	nr	3.05	2.5	3.13
adaptor, imperial to metric, 22mm x 3/4in	nr	0.89	2.5	0.91
reducing elbow, 22 x 15mm	nr	3.94	2.5	4.04
street elbow	nr	1.88	2.5	1.93
male elbow, 22mm x 3/4in	nr	3.39	2.5	3.47
female elbow, 22mm x 3/4in	nr	3.07	2.5	3.15
reducing female elbow, 22mm x 1/2in	nr	3.27	2.5	3.35
backplate elbow, 22mm x 3/4in	nr	5.74	2.5	5.88
flanged bend, 22mm x 3/4in	nr	4.93	2.5	5.05
overflow bend, 22mm x 3/4in	nr	6.18	2.5	6.33
slow bend	nr	2.50	2.5	2.56
return bend	nr	6.18	2.5	6.33
obtuse elbow	nr	1.97	2.5	2.02
tee, reduced branch, largest end, 22mm	nr	3.63	2.5	3.72
tee, both ends reduced, largest end, 22mm	nr	4.50	2.5	4.61
corner tee	nr	6.90	2.5	7.07
sweep tee, 90 degrees	nr	4.42	2.5	4.53
sweep tee, reduced branch, largest end, 22mm	nr	4.65	2.5	4.77
offset tee, 22mm equal	nr	6.14	2.5	6.29
double sweep tee, 22mm	nr	6.63	2.5	6.80
cross equal tee, 22mm	nr	7.21	2.5	7.39
stop end, 22mm	nr	2.43	2.5	2.49
tap connector, 22mm x 1/2in	nr	4.70	2.5	4.82
bent tap connector, 22mm x 1/2in	nr	5.72	2.5	5.86
bent tap connector, 22mm x 3/4in	nr	2.93	2.5	3.00
bent union adaptor, 22mm x 1in	nr	3.85	2.5	3.95

Capillary fittings (cont'd)	Unit	Material supply (£)	Material waste (%)	Total (£)
bent male union connector, 22mm x 3/4in	nr	6.50	2.5	6.66
bent female union connector, 22mm x 3/4in	nr	6.50	2.5	6.66
28mm fittings				
straight coupling	nr	0.73	2.5	0.75
reduced coupling, 28 x 15mm	nr	2.37	2.5	2.43
reduced coupling, 28 x 22mm	nr	1.44	2.5	1.48
adaptor coupling, imperial to metric, 1in x 28mm	nr	1.61	2.5	1.65
flanged connector	nr	21.82	2.5	22.37
straight female connector, 28mm x 1in	nr	3.63	2.5	3.72
straight male connector, 28mm x 1in	nr	3.07	2.5	3.15
male reducing connector, 28mm x 3/4in	nr	4.65	2.5	4.77
lead connector	nr	1.87	2.5	1.92
tank connector, 28mm x 1in	nr	5.50	2.5	5.64
tank connector, 28mm x 1 1/4in	nr	13.19	2.5	13.52
reducer, 28 x 15mm	nr	1.29	2.5	1.32
reducer, 28 x 22mm	nr	1.08	2.5	1.11
female adaptor, 28mm x 1in	nr	4.65	2.5	4.77
male adaptor, 28mm x 1in	nr	4.65	2.5	4.77
adaptor, imperial to metric, 28mm x 1in	nr	1.52	2.5	1.56
reducing elbow, 28 x 22mm	nr	6.30	2.5	6.46
street elbow	nr	3.42	2.5	3.51
male elbow, 28mm x 1in	nr	5.07	2.5	5.20
female elbow, 28mm x 1in	nr	4.81	2.5	4.93
slow bend	nr	4.16	2.5	4.26
return bend	nr	9.14	2.5	9.37
obtuse elbow	nr	3.44	2.5	3.53
equal tee, 28mm	nr	3.23	2.5	3.31
tee, reduced branch, largest end, 28mm	nr	3.47	2.5	3.56
tee, one end and one branch reduced, largest end, 28mm	nr	4.74	2.5	4.86
tee, both ends reduced, largest end, 28mm	nr	4.50	2.5	4.61
sweep tee, 90 degrees	nr	9.16	2.5	9.39

	Unit	Material supply (£)	Material waste (%)	Total (£)
sweep tee, reduced branch largest end, 28mm	nr	7.88	2.5	8.08
double sweep tee	nr	10.08	2.5	10.33
cross equal tee	nr	10.33	2.5	10.59
stop end, 28mm	nr	2.74	2.5	2.81
bent union adaptor, 28mm x 1 1/4in	nr	6.05	2.5	6.20
bent male union connector, 28mm x 1in	nr	9.48	2.5	9.72
bent female union connector, 28mm x 1in	nr	9.48	2.5	9.72
35mm fittings				
straight coupling	nr	1.93	0.0	1.93
reduced coupling, 35 x 22mm	nr	3.48	2.5	3.57
reduced coupling, 35 x 28mm	nr	4.31	2.5	4.42
adaptor coupling, imperial to metric, 1 1/4in x 35mm	nr	3.71	2.5	3.80
flanged connector,	nr	27.63	2.5	28.32
straight female connector, 35mm x 1 1/4in	nr	5.65	2.5	5.79
straight male connector, 35mm x 1 1/4in	nr	5.54	2.5	5.68
tank connector, 35mm x 1 1/4in	nr	7.52	2.5	7.71
reducer, 35 x 15mm	nr	2.59	2.5	2.65
reducer, 35 x 22mm	nr	2.48	2.5	2.54
reducer, 35 x 28mm	nr	2.48	2.5	2.54
female adaptor, 35mm x 1 1/4in	nr	6.86	2.5	7.03
male adaptor, 35mm x 1 1/4in	nr	6.86	2.5	7.03
adaptor, imperial to metric, 1 1/4in x 35mm	nr	1.88	2.5	1.93
elbow	nr	4.03	2.5	4.13
street elbow	nr	6.37	2.5	6.53
male elbow, 35mm x 1 1/4in	nr	7.05	2.5	7.23
female elbow, 35mm x 1 1/4in	nr	7.68	2.5	7.87
obtuse elbow	nr	8.01	2.5	8.21
equal tee, 35mm	nr	7.30	2.5	7.48
tee, reduced branch, largest end, 35mm	nr	7.58	2.5	7.77
tee, one end and one branch reduced, largest end, 35mm	nr	9.34	2.5	9.57
tee, both ends reduced, largest end, 35mm	nr	13.09	2.5	13.42

Capillary fittings (cont'd)	Unit	Material supply (£)	Material waste (%)	Total (£)
sweep tee, 90 degrees, 35mm	nr	13.78	2.5	14.12
sweep tee, reduced branch largest end, 35mm	nr	13.78	2.5	14.12
stop end, 35mm	nr	4.27	2.5	4.38
bent union adaptor, 35mm x 1 1/2in	nr	9.79	2.5	10.03
bent male union connector, 35mm x 1 1/4in	nr	14.77	2.5	15.14
bent female union connector, 35mm x 1 1/4in	nr	14.77	2.5	15.14
42mm fittings				
straight coupling	nr	2.86	2.5	2.93
reduced coupling, 42 x 28mm	nr	5.50	2.5	5.64
reduced coupling, 42 x 35mm	nr	5.50	2.5	5.64
adaptor coupling, imperial to metric, 1 1/2in x 42mm	nr	4.71	2.5	4.83
flanged connector	nr	33.02	2.5	33.85
straight female connector, 42mm x 1 1/2in	nr	7.07	2.5	7.25
straight male connector, 42mm x 1 1/2in	nr	7.21	2.5	7.39
tank connector, 42mm x 1 1/2in	nr	9.86	2.5	10.11
reducer, 42 x 15mm	nr	3.80	2.5	3.90
reducer, 42 x 22mm	nr	4.33	2.5	4.44
reducer, 42 x 28mm	nr	4.33	2.5	4.44
reducer, 42 x 35mm	nr	3.80	2.5	3.90
female adaptor, 42mm x 1 1/2in	nr	10.54	2.5	10.80
male adaptor, 42mm x 1 1/2in	nr	10.54	2.5	10.80
adaptor, imperial to metric, 1 1/2in x 42mm	nr	2.49	2.5	2.55
elbow	nr	6.46	2.5	6.62
street elbow	nr	9.42	2.5	9.66
male elbow, 42mm x 1 1/2in	nr	8.44	2.5	8.65
obtuse elbow	nr	12.48	2.5	12.79
equal tee, 42mm	nr	10.97	2.5	11.24
tee, reduced branch, largest end, 42mm	nr	13.37	2.5	13.70
tee, one end and one branch reduced, largest end, 42mm	nr	13.37	2.5	13.70
tee, both ends reduced, largest end, 42mm	nr	21.14	2.5	21.67
sweep tee, 90 degrees, 42mm	nr	19.95	2.5	20.45

	Unit	Material supply (£)	Material waste (%)	Total (£)
stop end, 42mm	nr	5.67	2.5	5.81
bent male union connector, 42mm x 2in	nr	25.52	2.5	26.16
bent female union connector, 42mm x 1 1/2in	nr	25.52	2.5	26.16
54mm fittings				
straight coupling	nr	6.70	2.5	6.87
reduced coupling, 54 x 35mm	nr	8.05	2.5	8.25
reduced coupling, 54 x 42mm	nr	8.05	2.5	8.25
flanged connector	nr	49.93	2.5	51.18
straight female connector, 54mm x 2in	nr	12.08	2.5	12.38
straight male connector, 54mm x 2in	nr	10.97	2.5	11.24
tank connector, 54mm x 2in	nr	15.04	2.5	15.42
reducer, 54 x 15mm	nr	7.80	2.5	8.00
reducer, 54 x 22mm	nr	7.80	2.5	8.00
reducer, 54 x 28mm	nr	6.79	2.5	6.96
reducer, 54 x 35mm	nr	6.79	2.5	6.96
reducer, 54 x 42mm	nr	6.79	2.5	6.96
female adaptor, 54mm x 2in	nr	12.46	2.5	12.77
male adaptor, 54mm x 2in	nr	12.46	2.5	12.77
adaptor, imperial to metric, 54mm x 2in	nr	5.08	2.5	5.21
elbow	nr	14.71	2.5	15.08
male elbow, 54mm x 2in	nr	14.71	2.5	15.08
obtuse elbow	nr	16.45	2.5	16.86
equal tee, 54mm	nr	20.61	2.5	21.13
tee, reduced branch, largest end, 54mm	nr	21.14	2.5	21.67
tee, one end and one branch reduced, largest end, 54mm	nr	21.14	2.5	21.67
sweep tee, 90 degrees, 54mm	nr	24.87	2.5	25.49
stop end, 54mm	nr	8.56	2.5	8.77
bent union adaptor, 54mm x 2 1/2in	nr	27.17	2.5	27.85
bent male union connector, 54mm x 2in	nr	40.66	2.5	41.68
bent female union connector, 54mm x 2in	nr	40.66	2.5	41.68

	Unit	Material supply (£)	Material waste (%)	Total (£)
Stopcocks				
Gunmetal stopcock with brass headwork, copper x copper				
15mm	nr	2.93	2.5	3.00
22mm	nr	5.47	2.5	5.61
28mm	nr	15.55	2.5	15.94
Dezincification-resistant stopcock, copper x copper				
15mm	nr	7.36	2.5	7.54
22mm	nr	12.74	2.5	13.06
28mm	nr	21.22	2.5	21.75
Gunmetal stopcock, copper x copper				
35mm	nr	33.30	2.5	34.13
42mm	nr	44.23	2.5	45.34
54mm	nr	66.10	2.5	67.75
Gunmetal lockshield stopcock with brass headwork, copper x copper				
15mm	nr	7.18	2.5	7.36
22mm	nr	10.30	2.5	10.56
28mm	nr	18.27	2.5	18.73
Dezincification-resistant lock-shield stopcock, copper x copper				
15mm	nr	9.67	2.5	9.91
22mm	nr	14.41	2.5	14.77
28mm	nr	23.96	2.5	24.56
Gunmetal stopcock, with brass headwork, parallel main thread x copper				
15mm x 1/2in	nr	7.71	2.5	7.90

MATERIAL COSTS

	Unit	Material supply (£)	Material waste (%)	Total (£)
Gunmetal stopcock with brass headwork, parallel main thread with backnut x copper				
15mm x 1/2in	nr	8.42	2.5	8.63
Gunmetal stopcock, with brass headwork, lead x copper				
15mm	nr	6.72	2.5	6.89
Gunmetal union stopcock with brass headwork, copper x copper union				
15mm	nr	8.24	2.5	8.45
22mm	nr	12.14	2.5	12.44
28mm	nr	19.86	2.5	20.36
Gunmetal union lockshield stopcock with brass headwork, copper x copper union				
15mm	nr	8.24	2.5	8.45
22mm	nr	12.14	2.5	12.44
28mm	nr	19.86	2.5	20.36
Gunmetal double union stopcock with brass headwork, copper x copper union				
15mm	nr	9.46	2.5	9.70
22mm	nr	13.38	2.5	13.71
28mm	nr	23.77	2.5	24.36
Dezincification-resistant double union stopcock, copper x copper				
15mm	nr	15.46	2.5	15.85
22mm	nr	19.04	2.5	19.52
28mm	nr	35.18	2.5	36.06

Stopcocks (cont'd)	Unit	Material supply (£)	Material waste (%)	Total (£)
Gunmetal double union stopcock, copper x copper				
35mm	nr	58.70	2.5	60.17
42mm	nr	80.51	2.5	82.52
54mm	nr	126.71	2.5	129.88
Gunmetal double union lockshield stopcock, with brass headwork, copper x copper				
15mm	nr	10.63	2.5	10.90
22mm	nr	21.49	2.5	22.03
28mm	nr	26.58	2.5	27.24
Dezincification-resistant double union lockshield stopcock, copper x copper				
15mm	nr	14.42	2.5	14.78
22mm	nr	20.56	2.5	21.07
28mm	nr	37.97	2.5	38.92
Gunmetal stopcock with easy-clean cover and brass headwork, copper x copper				
15mm	nr	7.55	2.5	7.74
22mm	nr	10.71	2.5	10.98
28mm	nr	18.84	2.5	19.31
Gunmetal union stopcock, with easy-clean cover and brass headwork, copper x copper union				
15mm	nr	9.20	2.5	9.43
22mm	nr	14.00	2.5	14.35
28mm	nr	22.29	2.5	22.85

MATERIAL COSTS

	Unit	Material supply (£)	Material waste (%)	Total (£)
Gunmetal double union stopcock, with easy-clean cover and brass headwork				
15mm	nr	11.02	2.5	11.30
22mm	nr	15.22	2.5	15.60
28mm	nr	26.77	2.5	27.44
Dezincification-resistant double union stopcock, with easy-clean cover, copper x copper				
15mm	nr	16.58	2.5	16.99
22mm	nr	20.66	2.5	21.18
28mm	nr	38.39	2.5	39.35
Gunmetal double union stopcock with easy-clean cover, copper x copper				
35mm	nr	63.60	2.5	65.19
42mm	nr	88.12	2.5	90.32
54mm	nr	141.71	2.5	145.25
Combined gunmetal stopcock and draincock with brass headwork, copper x copper union				
15mm	nr	16.29	2.5	16.70
22mm	nr	20.05	2.5	20.55
Combined dezincification-resistant stopcock and draincock, copper x copper union				
15mm	nr	21.58	2.5	22.12

	Unit	Material supply (£)	Material waste (%)	Total (£)
Gate valves to BS2871.1 lead-free pre-soldered joints				
Gunmetal fullway gate valve, copper x copper				
15mm	nr	9.20	2.5	9.43
22mm	nr	10.64	2.5	10.91
28mm	nr	14.81	2.5	15.18
35mm	nr	33.01	2.5	33.84
42mm	nr	41.31	2.5	42.34
54mm	nr	59.88	2.5	61.38
Gunmetal lockshield gate valve, copper x copper				
15mm	nr	9.46	2.5	9.70
22mm	nr	11.06	2.5	11.34
28mm	nr	17.25	2.5	17.68
35mm	nr	33.30	2.5	34.13
42mm	nr	41.31	2.5	42.34
54mm	nr	59.88	2.5	61.38
Plug cocks				
Nited				
12mm	nr	5.83	2.5	5.98
15mm	nr	6.81	2.5	6.98
22mm	nr	10.23	2.5	10.49
Nited with drop head				
12mm	nr	6.44	2.5	6.60
15mm	nr	7.73	2.5	7.92
22mm	nr	10.58	2.5	10.84
Nited with union outlet				
15mm	nr	7.83	2.5	8.03
22mm	nr	12.01	2.5	12.31

MATERIAL COSTS

	Unit	Material supply (£)	Material waste (%)	Total (£)
Plug stopcock, copper x copper				
15mm	nr	7.88	2.5	8.08
22mm	nr	12.01	2.5	12.31
Plug stopcock, copper x male thread with backnut				
15mm x 1/2in	nr	8.24	2.5	8.45
22mm x 3/4in	nr	12.01	2.5	12.31
Manifolds				
Linear flow capillary socket inlet x four soft copper tube outlets				
22mm x four 8mm	nr	7.07	2.5	7.25
Linear flow capillary socket inlet x four soft copper tube outlets				
22mm x four 10mm	nr	9.08	2.5	9.31
Linear flow capillary socket inlet x six soft copper tube outlets				
22mm x six 8mm	nr	10.12	2.5	10.37
Side entry, capillary socket inlet x four compression outlets				
22mm x four 8mm	nr	8.22	2.5	8.43
Side entry, capillary socket inlet x six compression outlets				
22mm x six 8mm	nr	9.80	2.5	10.05
Side entry, capillary socket inlet x two compression outlets				
22mm x two 10mm	nr	5.37	2.5	5.50

	Unit	Material supply (£)	Material waste (%)	Total (£)
Side entry, capillary socket inlet x four compression outlets				
22mm x four 10mm	nr	8.60	2.5	8.82
Valves				
Mistral radiator valve, plain finish, compression inlet x taper male union outlet				
8mm x 1/2in	nr	4.79	2.5	4.91
10mm x 1/2in	nr	4.79	2.5	4.91
15mm x 1/2in	nr	4.79	2.5	4.91
Mistral radiator valve, chromium plated finish, compression inlet x taper male union outlet				
8mm x 1/2in	nr	5.12	2.5	5.25
10mm x 1/2in	nr	5.12	2.5	5.25
15mm x 1/2in	nr	5.12	2.5	5.25
Maxitwin radiator valve, plain finish, compression connection x taper male union outlet				
two 8mm x 1/2in	nr	11.70	2.5	11.99
two 10mm x 1/2in	nr	12.78	2.5	13.10
Maxitwin radiator valve, chromium plated finish, compression connection x taper male union outlet				
two 8mm x 1/2in	nr	14.21	2.5	14.57
two 10mm x 1/2in	nr	15.60	2.5	15.99

	Unit	Material supply (£)	Material waste (%)	Total (£)
Dezincification-resistant compression fittings				
15mm				
straight coupling, 15mm	nr	1.37	2.5	1.40
male coupling, 15 x 1/2in	nr	1.15	2.5	1.18
male coupling, 15 x 3/4in	nr	2.59	2.5	2.65
male coupling, BSP parallel thread 15mm x 1/2in	nr	1.17	2.5	1.20
tank coupling with back nut, 15mm x 1/2in	nr	3.10	2.5	3.18
female coupling, 15mm x 1/2in	nr	1.32	2.5	1.35
female coupling, 15mm x 3/4in	nr	2.30	2.5	2.36
elbow, 15mm	nr	1.60	2.5	1.64
male elbow, 15mm x 1/2in	nr	1.96	2.5	2.01
male elbow, BSP parallel thread, 15mm x 1/2in	nr	1.95	2.5	2.00
female elbow, 15mm x 1/2in	nr	2.15	2.5	2.20
female elbow, 15mm x 3/4in	nr	4.16	2.5	4.26
female wall elbow, 15mm x 1/2in	nr	3.20	2.5	3.28
tee, 15mm	nr	2.20	2.5	2.26
tee with reduced branch, 15 x 15 x 12mm	nr	3.90	2.5	4.00
tee, end reduced, 15 x 12 x 15mm	nr	2.96	2.5	3.03
tee, end and branch reduced, 15 x 12 x 12mm	nr	3.36	2.5	3.44
tee, both ends reduced, 15 x 15 x 18mm	nr	5.53	2.5	5.67
tee, both ends reduced, 15 x 15 x 22mm	nr	4.38	2.5	4.49
male tee, 15mm x 1/2in x 15mm	nr	4.35	2.5	4.46
male tee, 15mm x 15mm x 1/2in	nr	4.62	2.5	4.74
cross tee, 15mm	nr	5.43	2.5	5.57
straight swivel connector, 15mm x 1/2in	nr	2.43	2.5	2.49
straight swivel connector, 15mm x 3/4in	nr	3.51	2.5	3.60
bent swivel connector, 15mm x 1/2in	nr	2.93	2.5	3.00
female tee, 15mm x 15mm x 1/2in	nr	3.46	2.5	3.55
stop end, 15mm	nr	1.60	2.5	1.64

DZR fittings (cont'd)	Unit	Material supply (£)	Material waste (%)	Total (£)
18mm				
straight coupling, 18mm	nr	2.52	2.5	2.58
male coupling, 18mm x 1/2in	nr	2.13	2.5	2.18
female coupling, 18mm x 1/2in	nr	2.24	2.5	2.30
elbow, 18mm	nr	2.76	2.5	2.83
male elbow, 18mm x 1/2in	nr	2.76	2.5	2.83
female elbow, 18mm x 1/2in	nr	3.30	2.5	3.38
female wall elbow, 18mm x 1/2in	nr	6.22	2.5	6.38
tee, 18mm	nr	3.87	2.5	3.97
tee with reduced branch, 18 x 18 x 12mm	nr	5.53	2.5	5.67
tee with reduced branch, 18 x 18 x 15mm	nr	3.90	2.5	4.00
tee, end reduced, 18 x 12 x 18mm	nr	5.22	2.5	5.35
tee, end reduced, 18 x 15 x 18mm	nr	5.53	2.5	5.67
tee, end and branch reduced, 18 x 12 x 12mm	nr	5.22	2.5	5.35
tee, end and branch reduced, 18 x 15 x 12mm	nr	5.22	2.5	5.35
tee, end and branch reduced, 18 x 15 x 15mm	nr	5.53	2.5	5.67
female adaptor, 18mm x 1/2in	nr	2.66	2.5	2.73
reducing set, 18 x12mm	nr	0.94	2.5	0.96
reducing set, 18 x 15mm	nr	0.94	2.5	0.96
22mm				
straight coupling, 22mm	nr	2.12	2.5	2.17
male coupling, 22mm x 1/2in	nr	2.29	2.5	2.35
male coupling, 22mm x 3/4in	nr	1.71	2.5	1.75
male coupling, 22mm x 1in	nr	2.60	2.5	2.67
male coupling, BSP parallel thread, 22mm x 1in	nr	2.60	2.5	2.67
tank coupling with back nut, 22mm x 3/4in	nr	3.33	2.5	3.41
tank coupling with back nut, 22mm x 1in	nr	5.84	2.5	5.99
female coupling, 22mm x 1/2in	nr	2.36	2.5	2.42
female coupling, 22mm x 3/4in	nr	1.99	2.5	2.04
female coupling, 22mm x 1in	nr	3.57	2.5	3.66
elbow, 22mm	nr	2.59	2.5	2.65
male elbow, 22mm x 3/4in	nr	2.28	2.5	2.34
male elbow, 22mm x 1in	nr	4.70	2.5	4.82

MATERIAL COSTS

	Unit	Material supply (£)	Material waste (%)	Total (£)
male elbow, BSP parallel thread, 22mm x 3/4in	nr	2.78	2.5	2.85
male elbow, BSP parallel thread, 22mm x 1in	nr	4.70	2.5	4.82
female elbow, 22mm x 3/4in	nr	3.10	2.5	3.18
female elbow, 22mm x 1in	nr	5.36	2.5	5.49
female wall elbow, 22mm x 3/4in	nr	5.87	2.5	6.02
tee, 22mm	nr	3.72	2.5	3.81
tee with reduced branch, 22 x 22 x 12mm	nr	5.25	2.5	5.38
tee with reduced branch, 22 x 22 x 15mm	nr	4.58	2.5	4.69
tee, end reduced, 22 x 12 x 22mm	nr	5.12	2.5	5.25
tee, end reduced, 22 x 15 x 22mm	nr	4.58	2.5	4.69
tee, end and branch reduced, 22 x 15 x 15mm	nr	4.58	2.5	4.69
tee, both ends reduced, 22 x 22 x 28mm	nr	7.99	2.5	8.19
male tee, 22mm x 3/4in	nr	6.86	2.5	7.03
male tee, 22mm x 22mm x 3/4in	nr	6.23	2.5	6.39
straight swivel connector, 22mm x 3/4in	nr	4.12	2.5	4.22
bent swivel connector, 22mm x 3/4in	nr	4.75	2.5	4.87
female tee, 22mm x 22mm x 1/2in	nr	4.78	2.5	4.90
female tee, 22mm x 22mm x 3/4in	nr	4.87	2.5	4.99
male adaptor, 22mm x 1/2in	nr	3.14	2.5	3.22
male adaptor, 22mm x 3/4in	nr	3.14	2.5	3.22
female adaptor, 22mm x 1/2in	nr	3.14	2.5	3.22
female adaptor, 22mm x 3/4in	nr	2.86	2.5	2.93
reducing set, 22 x 12mm	nr	1.69	2.5	1.73
reducing set, 22 x 15mm	nr	1.27	2.5	1.30
reducing set, 22 x 18mm	nr	0.98	2.5	1.00
stop end, 22mm	nr	2.40	2.5	2.46
28mm				
straight coupling, 28mm	nr	4.07	2.5	4.17
male coupling, 28mm x 3/4in	nr	3.95	2.5	4.05
male coupling, 28mm x 1in	nr	3.09	2.5	3.17
male coupling, BSP parallel thread, 28mm x 1in	nr	3.05	2.5	3.13
tank coupling with back nut, 28mm x 1in	nr	6.09	2.5	6.24

DZR fittings (cont'd)	Unit	Material supply (£)	Material waste (%)	Total (£)
female coupling, 28mm x 1in	nr	3.32	2.5	3.40
elbow, 28mm	nr	5.86	2.5	6.01
male elbow, 28mm x 1in	nr	4.00	2.5	4.10
male elbow, BSP parallel thread, 28mm x 1in	nr	4.00	2.5	4.10
female elbow, 28mm x 1in	nr	4.88	2.5	5.00
tee, 28mm	nr	7.39	2.5	7.57
tee with reduced branch, 28 x 28 x 12mm	nr	4.98	2.5	5.10
tee with reduced branch, 28 x 28 x 15mm	nr	7.56	2.5	7.75
tee with reduced branch, 28 x 28 x 22mm	nr	7.56	2.5	7.75
tee, end reduced, 28 x 15 x 28mm	nr	7.99	2.5	8.19
tee, end reduced, 28 x 22 x 28mm	nr	7.56	2.5	7.75
tee, end and branch reduced, 28 x 22 x 15mm	nr	7.99	2.5	8.19
tee, end and branch reduced, 28 x 22 x 22mm	nr	7.99	2.5	8.19
male tee, 28mm x 1in x 28mm	nr	11.66	2.5	11.95
female tee 28mm x 28mm x 1/2in	nr	7.19	2.5	7.37
female tee 28mm x 28mm x 1in	nr	8.70	2.5	8.92
male adaptor, 28mm x 1in	nr	5.15	2.5	5.28
female adaptor, 28mm x 1in	nr	5.15	2.5	5.28
reducing set, 28 x 12mm	nr	2.03	2.5	2.08
reducing set, 28 x 15mm	nr	1.90	2.5	1.95
reducing set, 28 x 22mm	nr	1.90	2.5	1.95
stop end, 28mm	nr	4.00	2.5	4.10
35mm				
straight coupling, 35mm	nr	7.90	2.5	8.10
male coupling, 35mm x 1in	nr	7.67	2.5	7.86
male coupling, 35mm x 1 1/4in	nr	6.18	2.5	6.33
tank coupling with back nut, 35mm x 1 1/4in	nr	6.84	2.5	7.01
female coupling, 35mm x 1 1/4in	nr	6.78	2.5	6.95
elbow, 35mm	nr	10.26	2.5	10.52
male elbow, 35mm x 1 1/4in	nr	9.94	2.5	10.19
female elbow, 35mm x 1 1/4in	nr	9.47	2.5	9.71
tee, 35mm	nr	13.55	2.5	13.89
tee with reduced branch, 35 x 35 x 15mm	nr	13.55	2.5	13.89

	Unit	Material supply (£)	Material waste (%)	Total (£)
tee with reduced branch, 35 x 35 x 22mm	nr	13.55	2.5	13.89
tee with reduced branch, 35 x 35 x 28mm	nr	13.55	2.5	13.89
reducing set, 35 x 15mm	nr	3.24	2.5	3.32
reducing set, 35 x 22mm	nr	3.24	2.5	3.32
reducing set, 35 x 28mm	nr	3.24	2.5	3.32
tank coupling, 35mm	nr	10.10	2.5	10.35
stop end, 35mm	nr	5.85	2.5	6.00
42mm				
straight coupling, 42mm	nr	10.15	2.5	10.40
male coupling, 42mm x 1 1/2in	nr	9.00	2.5	9.23
tank coupling with back nut, 42mm x 1 1/2in	nr	9.48	2.5	9.72
female coupling, 42mm x 1 1/2in	nr	9.20	2.5	9.43
elbow, 42mm	nr	14.43	2.5	14.79
male elbow, 42mm x 1 1/2in	nr	14.21	2.5	14.57
female elbow, 42mm x 1 1/2in	nr	14.21	2.5	14.57
tee, 42mm	nr	22.56	2.5	23.12
tee with reduced branch, 42 x 42 x 22mm	nr	21.13	2.5	21.66
tee with reduced branch, 42 x 42 x 28mm	nr	21.13	2.5	21.66
reducing set, 42 x 15mm	nr	5.91	2.5	6.06
reducing set, 42 x 22mm	nr	5.11	2.5	5.24
reducing set, 42 x 28mm	nr	5.11	2.5	5.24
reducing set, 42 x 35mm	nr	5.11	2.5	5.24
stop end, 42mm	nr	10.17	2.5	10.42
54mm				
straight coupling, 54mm	nr	15.16	2.5	15.54
male coupling, 54mm x 2in	nr	12.97	2.5	13.29
tank coupling with back nut, 54mm x 2in	nr	14.60	2.5	14.97
female coupling, 54mm x 2in	nr	13.33	2.5	13.66
elbow, 54mm	nr	20.27	2.5	20.78
male elbow, 54mm x 2in	nr	21.09	2.5	21.62
tee, 54mm	nr	35.35	2.5	36.23
tee with reduced branch, 54 x 54 x 28mm	nr	35.35	2.5	36.23
reducing set, 54 x 22mm	nr	9.81	2.5	10.06

DZR fittings (cont'd)	Unit	Material supply (£)	Material waste (%)	Total (£)
reducing set, 54 x 28mm	nr	9.81	2.5	10.06
reducing set, 54 x 35mm	nr	8.54	2.5	8.75
reducing set, 54 x 42mm	nr	8.54	2.5	8.75

Brass compression fittings to BS864

15mm

straight coupling	nr	0.92	2.5	0.94
male coupling, extended thread, 15mm x 1/2in	nr	3.10	2.5	3.18
lead coupling	nr	2.31	2.5	2.37
bent union radiator plain, 15mm x 1/2in	nr	3.70	2.5	3.79
air release elbow, 15mm	nr	4.90	2.5	5.02
slow bend	nr	4.94	2.5	5.06
male elbow, 15mm x 1/2in	nr	1.39	2.5	1.42
male elbow, 15mm x 3/4in	nr	4.10	2.5	4.20
male elbow, BSP parallel thread, 15mm x 1/2in	nr	1.42	2.5	1.46
female elbow	nr	1.91	2.5	1.96
female wall elbow, 15mm x 1/2in	nr	2.83	2.5	2.90
female wall elbow, 15mm x 3/4in	nr	6.11	2.5	6.26
female tee with back plate, 15 x 15 x 15mm	nr	5.00	2.5	5.13
offset tee, 15 x 15 x 15mm	nr	8.41	2.5	8.62
male adaptor, 15mm x 1/2in	nr	1.58	2.5	1.62
reducing set, 15 x 8mm	nr	1.11	2.5	1.14
reducing set, 15 x 10mm	nr	1.11	2.5	1.14
reducing set, 15 x 12mm	nr	1.12	2.5	1.15
sweep tee, 15mm	nr	8.39	2.5	8.60
corner tee, 15 x 15 x 15mm	nr	8.60	2.5	8.82

22mm

straight coupling	nr	1.54	2.5	1.58
male coupling, extended thread, 22mm x 3/4in	nr	3.94	2.5	4.04
female coupling, 22mm x 1in	nr	2.52	2.5	2.58
lead coupling	nr	3.27	2.5	3.35
air release elbow, 22mm	nr	6.19	2.5	6.34
slow bend, 22mm	nr	8.10	2.5	8.30
male elbow, 22mm x 3/4in	nr	1.78	2.5	1.82

	Unit	Material supply (£)	Material waste (%)	Total (£)
male elbow, 22mm x 1in	nr	3.15	2.5	3.23
male elbow, BSP parallel thread, 22mm x 3/4in	nr	1.83	2.5	1.88
male elbow, BSP parallel thread, 22mm x 1in	nr	3.15	2.5	3.23
female elbow, 22mm x 3/4in	nr	2.78	2.5	2.85
female elbow, 22mm x 1in	nr	4.64	2.5	4.76
bent cylinder connectors, 22mm x 1in	nr	6.98	2.5	7.15
tee with reduced branch, 22 x 22 x 15mm	nr	4.30	2.5	4.41
tee end reduced, 22 x 15 x 22mm	nr	4.39	2.5	4.50
tee, end and branch reduced, 22 x 15 x 15mm	nr	4.41	2.5	4.52
straight swivel connector, 22mm x 1/2in	nr	3.70	2.5	3.79
bent swivel connector, 22mm x 1/2in	nr	5.23	2.5	5.36
offset tee, 22 x 22 x 15mm	nr	12.39	2.5	12.70
tank coupling, extended thread, 22mm	nr	5.17	2.5	5.30
sweep tee, 22mm	nr	9.09	2.5	9.32
28mm				
straight coupling	nr	3.77	2.5	3.86
male coupling, 28mm x 1 1/4in	nr	7.51	2.5	7.70
female coupling, 28mm x 1in	nr	4.10	2.5	4.20
female coupling, 28mm x 1in	nr	2.60	2.5	2.67
slow bend	nr	10.90	2.5	11.17
male elbow, 28mm x 1in	nr	3.81	2.5	3.91
male elbow, BSP parallel thread, 28mm x 1in	nr	3.81	2.5	3.91
female elbow, 28mm x 1in	nr	4.62	2.5	4.74
tee, 28mm	nr	7.07	2.5	7.25
female tee, 28mm x 28mm x 3/4in	nr	9.06	2.5	9.29
tank coupling, 28mm	nr	4.71	2.5	4.83

	Unit	Material supply (£)	Material waste (%)	Total (£)
Stopcocks				
Gunmetal stopcocks with brass headwork, copper x copper				
15mm	nr	3.87	2.5	3.97
22mm	nr	6.62	2.5	6.79
28mm	nr	17.76	2.5	18.20
Gunmetal lockshield stopcock with brass headwork, copper x copper				
15mm	nr	8.46	2.5	8.67
22mm	nr	11.86	2.5	12.16
28mm	nr	22.88	2.5	23.45
Stopcock, dezincification-resistant, copper x copper				
15mm	nr	9.04	2.5	9.27
22mm	nr	14.84	2.5	15.21
28mm	nr	24.54	2.5	25.15
35mm	nr	45.15	2.5	46.28
42mm	nr	64.60	2.5	66.21
54mm	nr	87.99	2.5	90.19
Lockshield stopcock, dezincification-resistant, copper x copper				
15mm	nr	10.39	2.5	10.65
22mm	nr	16.28	2.5	16.69
Gunmetal stopcock with brass headwork, parallel male thread x copper				
15mm x 1/2in	nr	7.68	2.5	7.87
22mm x 3/4in	nr	10.67	2.5	10.94
Gunmetal stopcock with brass headwork, copper x lead inlet				
15mm	nr	7.51	2.5	7.70

MATERIAL COSTS

	Unit	Material supply (£)	Material waste (%)	Total (£)
Gunmetal stopcock with brass headwork, and easy-clean cover, copper x copper				
15mm	nr	9.30	2.5	9.53
22mm	nr	13.12	2.5	13.45
28mm	nr	23.50	2.5	24.09
Combined gunmetal stopcock and draincock, with brass headwork, copper x copper				
15mm	nr	13.01	2.5	13.34
22mm	nr	20.04	2.5	20.54
Combined stopcock and draincock, dezincification-resistant, copper x copper				
15mm	nr	16.77	2.5	17.19
22mm	nr	24.58	2.5	25.19
Gunmetal stopcock, soft copper x soft copper				
15mm	nr	14.31	2.5	14.67
28mm	nr	26.83	2.5	27.50
Gunmetal stopcock with brass headwork, polyethylene x polyethylene				
20mm	nr	10.49	2.5	10.75
25mm	nr	16.77	2.5	17.19
Stopcock dezincification-resistant polyethylene x polyethylene				
20mm	nr	14.08	2.5	14.43
25mm	nr	22.66	2.5	23.23
32mm	nr	30.50	2.5	31.26
50mm	nr	76.49	2.5	78.40
63mm	nr	107.88	2.5	110.58

Stopcocks (cont'd)	Unit	Material supply (£)	Material waste (%)	Total (£)
Gunmetal stopcock for underground use, dezincification-resistant, polyethylene x polyethylene				
20mm	nr	12.41	2.5	12.72
25mm	nr	19.95	2.5	20.45
Gunmetal stopcock with brass headwork, polyethylene x copper				
20 x 15mm	nr	9.72	2.5	9.96
25 x 15mm	nr	16.86	2.5	17.28
25 x 22mm	nr	17.99	2.5	18.44
Combined gunmetal stopcock and draincock with brass headwork, polyethylene x polyethylene				
20mm	nr	20.03	2.5	20.53
Combined gunmetal stopcock and draincock with brass headwork, polyethylene x copper				
20 x 15mm	nr	18.37	2.5	18.83
Combined gunmetal stopcock and draincock, dezincification-resistant, polyethylene x copper				
20 x 15mm	nr	27.14	2.5	27.82
25 x 15mm	nr	28.29	2.5	29.00
25 x 22mm	nr	32.58	2.5	33.39
Gunmetal stopcock with brass headwork, polyethylene x polyethylene				
3/8in	nr	10.32	2.5	10.58
1/2in	nr	10.59	2.5	10.85
3/4in	nr	15.97	2.5	16.37
1in	nr	23.08	2.5	23.66

	Unit	Material supply (£)	Material waste (%)	Total (£)
Gate valves to BS2871.1, brass compression fittings BS864				
Brass gate valve, copper x copper				
15mm	nr	6.32	2.5	6.48
22mm	nr	7.36	2.5	7.54
28mm	nr	12.83	2.5	13.15
Brass lockshield gate valve, copper x copper				
15mm	nr	6.37	2.5	6.53
22mm	nr	7.51	2.5	7.70
28mm	nr	13.61	2.5	13.95
Gunmetal fullway gate valve, copper x copper				
15mm	nr	10.95	2.5	11.22
22mm	nr	12.94	2.5	13.26
28mm	nr	17.58	2.5	18.02
35mm	nr	26.92	2.5	27.59
42mm	nr	35.37	2.5	36.25
54mm	nr	48.82	2.5	50.04
Gunmetal lockshield gate valve, copper x copper				
15mm	nr	11.28	2.5	11.56
22mm	nr	13.41	2.5	13.75
28mm	nr	20.52	2.5	21.03
35mm	nr	26.91	0.0	26.91
42mm	nr	37.07	0.0	37.07
54mm	nr	51.15	2.5	52.43
Check valves				
Dezincification-resistant alloy check valve with compression nuts				
15mm	nr	12.18	2.5	12.48
22mm	nr	13.01	2.5	13.34
28mm	nr	20.52	2.5	21.03

Check valves (cont'd)	Unit	Material supply (£)	Material waste (%)	Total (£)
Dezincification-resistant alloy check valve with female BSP parallel threads				
1/2in	nr	11.80	2.5	12.10
3/4in	nr	13.12	2.5	13.45
1in	nr	18.74	2.5	19.21
Drainage fittings to BS2879 Type 1				
Dezincification-resistant stopcock adaptor, copper x copper				
15mm	nr	6.16	2.5	6.31
Draw-off coupling, copper x copper				
15mm	nr	5.62	2.5	5.76
Dezincification-resistant draw-off coupling, copper x copper				
15mm	nr	7.74	2.5	7.93
22mm	nr	8.98	2.5	9.20
Dezincification-resistant male draw-off elbow, copper x parallel main thread				
15mm x 1/2in	nr	5.92	2.5	6.07
22mm x 3/4in	nr	6.65	2.5	6.82
28mm x 1in	nr	10.86	2.5	11.13
Brass draw-off elbow, copper to copper				
15mm	nr	5.75	2.5	5.89
22mm	nr	6.59	2.5	6.75
28mm	nr	10.16	2.5	10.41

MATERIAL COSTS

	Unit	Material supply (£)	Material waste (%)	Total (£)
Dezincification-resistant draw-off elbow, copper x copper				
15mm	nr	8.51	2.5	8.72
22mm	nr	10.50	2.5	10.76
Brass draincock, parallel thread connector				
1/2in	nr	1.72	2.5	1.76
Dezincification-resistant draincock, parallel threaded connector				
1/2in	nr	3.94	2.5	4.04
Servicing valves				
Straight pattern, copper x copper				
15 x 15mm	nr	3.52	2.5	3.61
22 x 22mm	nr	5.75	2.5	5.89
Straight pattern, copper x swivel end				
15mm x 1/2in	nr	3.69	2.5	3.78
Bent pattern, copper x swivel end				
15mm x 1/2in	nr	3.68	2.5	3.77
Appliance valves				
Chromium-plated stopvalve				
15mm x 3/4in	nr	11.73	2.5	12.02
Brass stopvalve				
15mm x 3/4in	nr	5.62	2.5	5.76

Appliance valves (cont'd)	Unit	Material supply (£)	Material waste (%)	Total (£)
Brass in-line valve				
15mm x 3/4in	nr	2.76	2.5	2.83
Brass elbow valve				
15mm x 3/4in	nr	2.76	2.5	2.83
Brass tee valve				
15mm x 15mm x 3/4in	nr	4.52	2.5	4.63
Cold water storage tanks				
Galvanized steel cistern, BS417 Part A with cover, reference				
SC10, 18 litres	nr	33.46	2.5	34.30
SC15, 36 litres	nr	37.91	2.5	38.86
SC20, 54 litres	nr	43.70	2.5	44.79
SC25, 68 litres	nr	45.37	2.5	46.50
SC30, 86 litres	nr	47.04	2.5	48.22
SC40, 114 litres	nr	53.52	2.5	54.86
SC50, 154 litres	nr	76.11	2.5	78.01
SC60, 191 litres	nr	80.82	2.5	82.84
SC70, 227 litres	nr	88.41	2.5	90.62
SC80, 260 litres	nr	92.92	2.5	95.24
SC100/2, 330 litres	nr	102.14	2.5	104.69
SC100/1, 320 litres	nr	103.67	2.5	106.26
SC125, 414 litres	nr	160.06	2.5	164.06
SC150, 480 litres	nr	174.93	2.5	179.30
SC200, 695 litres	nr	198.17	2.5	203.12
Plastic cistern, BS4213 with lid, reference				
PC4, 18 litres	nr	6.39	2.5	6.55
PC15, 68 litres	nr	22.45	2.5	23.01
PC20, 91 litres	nr	23.81	2.5	24.41
PC25, 114 litres	nr	30.34	2.5	31.10
PC40, 182 litres	nr	54.52	2.5	55.88
PC50, 227 litres	nr	44.21	2.5	45.32

	Unit	Material supply (£)	Material waste (%)	Total (£)
Hot water tanks				
Galvanized steel tank, BS417 with cover, reference				
T25/1, 36 litres	nr	108.97	2.5	111.69
T25/2, 36 litres	nr	110.66	2.5	113.43
T30/1, 86 litres	nr	116.75	2.5	119.67
T30/2, 86 litres	nr	118.19	2.5	121.14
T40, 114 litres	nr	143.14	2.5	146.72
Hot water copper cylinders				
Direct cylinder to BS699 grade 3, insulated, reference				
ref 5, 98 litres	nr	89.56	2.5	91.80
ref 7, 120 litres	nr	99.97	2.5	102.47
ref 8, 148 litres	nr	109.11	2.5	111.84
ref 9, 166 litres	nr	128.23	2.5	131.44
Indirect cylinder to BS699 grade 3, insulated, reference				
ref 2, 96 litres	nr	110.37	2.5	113.13
ref 3, 114 litres	nr	117.54	2.5	120.48
ref 7, 117 litres	nr	116.00	2.5	118.90
ref 8, 140 litres	nr	131.32	2.5	134.60
ref 9, 162 litres	nr	167.74	2.5	171.93
Fortic single-feed cylinder to BS3198, 4 connections, plain				
900 x 400mm, capacity 65 litres hot, 20 litres cold				
F1, direct	nr	136.53	2.5	139.94
F3, indirect coil	nr	170.55	2.5	174.81
F6, primatic indirect	nr	179.55	2.5	184.04
F8, primatic indirect	nr	191.50	2.5	196.29

Fortic single-feed (cont'd)	Unit	Material supply (£)	Material waste (%)	Total (£)
900 x 450mm, capacity 85 litres hot, 25 litres cold				
F1, direct	nr	140.60	2.5	144.11
F3, indirect coil	nr	176.32	2.5	180.73
F6, primatic indirect	nr	184.89	2.5	189.51
F8, primatic indirect	nr	195.44	2.5	200.33
1075 x 450mm, capacity 115 litres hot, 25 litres cold				
F1, direct	nr	154.66	2.5	158.53
F3, indirect coil	nr	191.78	2.5	196.57
F6, primatic indirect	nr	201.76	2.5	206.80
F8, primatic indirect	nr	212.73	2.5	218.05
1200 x 450mm, capacity 115 litres hot, 45 litres cold				
F1, direct	nr	164.65	2.5	168.77
F3, indirect coil	nr	201.62	2.5	206.66
F6, primatic indirect	nr	211.19	2.5	216.47
F8, primatic indirect	nr	223.00	2.5	228.58
1300 x 450mm, capacity 130 litres hot, 45 litres cold				
F1, direct	nr	192.20	2.5	197.01
F3, indirect coil	nr	225.39	2.5	231.02
F6, primatic indirect	nr	247.88	2.5	254.08
F8, primatic indirect	nr	260.11	2.5	266.61
1200 x 500mm, capacity 150 litres hot, 45 litres cold				
F1, direct	nr	233.82	2.5	239.67
F3, indirect coil	nr	267.57	2.5	274.26
F6, primatic indirect	nr	284.16	2.5	291.26
F8, primatic indirect	nr	295.41	2.5	302.80

	Unit	Material supply (£)	Material waste (%)	Total (£)
1400 x 500mm, capacity 115 litres hot, 115 litres cold				
F1, direct	nr	249.57	2.5	255.81
F3, indirect coil	nr	277.69	2.5	284.63
F6, primatic indirect	nr	298.50	2.5	305.96
F8, primatic indirect	nr	309.04	2.5	316.77
Insulation				
Preformed pipe lagging, fire-retardant foam, 22mm thick to pipe size				
15mm	m	0.68	5.0	0.71
22mm	m	0.84	5.0	0.88
28mm	m	1.00	5.0	1.05
50mm thick glass fibre filled insulating jacket, fixing bands to tank size				
445 x 305 x 300mm	nr	2.83	5.0	2.97
495 x 368 x 362mm	nr	5.15	5.0	5.41
630 x 450 x 420mm	nr	6.30	5.0	6.62
665 x 490 x 515mm	nr	6.90	5.0	7.25
700 x 540 x 535mm	nr	9.75	5.0	10.24
995 x 605 x 595mm	nr	12.10	5.0	12.71
60mm thick glass fibre filled insulating jacket, fixing bands, to cylinder 450mm diameter, height				
300mm	nr	6.40	5.0	6.72
450mm	nr	6.90	5.0	7.25
600mm	nr	7.80	5.0	8.19

	Unit	Labour hours	Net labour (£)	Net material (£)	O'heads /profit (£)	Total (£)
PIPEWORK						
Copper pipe to BS2871, Table W, pre-soldered capillary joints and fittings to BS864, clips at 1250mm maximum centres						
8mm diameter, to timber	m	0.20	1.92	1.29	0.48	3.69
8mm diameter, plugged and screwed	m	0.22	2.11	1.53	0.55	4.19
10mm diameter, to timber	m	0.20	1.92	1.38	0.49	3.80
10mm diameter, plugged and screwed	m	0.22	2.11	1.62	0.56	4.29
Copper pipe to BS2871, Table X, pre-soldered capillary joints and fittings to BS864, clips at 1250mm maximum centres						
15mm diameter, to timber	m	0.20	1.92	1.49	0.51	3.92
15mm diameter, plugged and screwed	m	0.22	2.11	1.73	0.58	4.42
22mm diameter, to timber	m	0.21	2.02	2.70	0.71	5.43
22mm diameter, plugged and screwed	m	0.23	2.21	2.95	0.77	5.93
28mm diameter, to timber	m	0.23	2.21	3.71	0.89	6.81
28mm diameter, plugged and screwed	m	0.25	2.40	3.95	0.95	7.30
35mm diameter, to timber	m	0.28	2.69	8.82	1.73	13.24
35mm diameter, plugged and screwed	m	0.30	2.88	9.06	1.79	13.73
42mm diameter, to timber	m	0.30	2.88	11.49	2.16	16.53
42mm diameter, plugged and screwed	m	0.32	3.07	11.74	2.22	17.03
54mm diameter, to timber	m	0.32	3.07	16.38	2.92	22.37
54mm diameter, plugged and screwed	m	0.34	3.26	16.63	2.98	22.87

	Unit	Labour hours	Net labour (£)	Net material (£)	O'heads /profit (£)	Total (£)
Copper pipe to BS2871, Table Z, lead-free pre-soldered capillary joints and fittings to BS864, two-piece clips at 1250mm maximum centres						
15mm diameter, to timber	m	0.20	1.92	1.44	0.50	3.86
15mm diameter, plugged and screwed	m	0.22	2.11	1.85	0.59	4.55
22mm diameter, to timber	m	0.21	2.02	2.39	0.66	5.07
22mm diameter, plugged and screwed	m	0.23	2.21	2.74	0.74	5.69
28mm diameter, to timber	m	0.23	2.21	3.71	0.89	6.81
28mm diameter, plugged and screwed	m	0.25	0.00	4.05	0.61	4.66
35mm diameter, to timber	m	0.28	2.69	7.60	1.54	11.83
35mm diameter, plugged and screwed	m	0.30	2.88	7.95	1.62	12.45
42mm diameter, to timber	m	0.30	2.88	10.08	1.94	14.90
42mm diameter, plugged and screwed	m	0.32	3.07	10.43	2.02	15.53
54mm diameter, to timber	m	0.32	3.07	14.35	2.61	20.03
54mm diameter, plugged and screwed	m	0.34	3.26	14.39	2.65	20.30
Copper pipe to BS2871, Table X, dezincification-resistant compression joints and fittings to BS864, two-piece clips 1250mm maximum centres						
15mm diameter, to timber	m	0.19	1.82	2.12	0.59	4.53
15mm diameter, plugged and screwed	m	0.21	2.02	2.36	0.66	5.04

Pipework (cont'd)	Unit	Labour hours	Net labour (£)	Net material (£)	O'heads /profit (£)	Total (£)
22mm diameter, to timber	m	0.20	1.92	3.94	0.88	6.74
22mm diameter, plugged and screwed	m	0.22	2.11	4.20	0.95	7.26
28mm diameter, to timber	m	0.22	2.11	5.49	1.14	8.74
28mm diameter, plugged and screwed	m	0.24	2.30	5.73	1.20	9.23
35mm diameter, to timber	m	0.27	2.59	12.02	2.19	16.80
35mm diameter, plugged and screwed	m	0.29	2.78	12.26	2.26	17.30
42mm diameter, to timber	m	0.29	2.78	14.95	2.66	20.39
42mm diameter, plugged and screwed	m	0.31	2.98	15.19	2.73	20.90
54mm diameter, to timber	m	0.31	2.98	20.51	3.52	27.01
54mm diameter, plugged and screwed	m	0.33	3.17	20.75	3.59	27.51

Copper pipe to BS2871, Table Z, dezincification-resistant compression joints and fittings to BS864, two-piece clips 1250mm maximum centres

Item	Unit	Labour hours	Net labour (£)	Net material (£)	O'heads /profit (£)	Total (£)
15mm diameter, to timber	m	0.19	1.82	2.07	0.58	4.47
15mm diameter, plugged and screwed	m	0.21	2.02	2.31	0.65	4.99
22mm diameter to timber	m	0.20	1.92	3.63	0.83	6.38
22mm diameter, plugged and screwed	m	0.22	2.11	3.89	0.90	6.90
28mm diameter, to timber	m	0.22	2.11	5.49	1.14	8.74
28mm diameter, plugged and screwed	m	0.24	2.30	5.73	1.20	9.23
35mm diameter, to timber	m	0.27	2.59	10.80	2.01	15.40

	Unit	Labour hours	Net labour (£)	Net material (£)	O'heads /profit (£)	Total (£)
35mm diameter, plugged and screwed	m	0.29	2.78	11.06	2.08	15.92
42mm diameter, to timber	m	0.29	2.78	13.62	2.46	18.86
42mm diameter, plugged and screwed	m	0.31	2.98	13.86	2.53	19.37
54mm diameter, to timber	m	0.31	2.98	18.17	3.17	24.32
54mm diameter, plugged and screwed	m	0.33	3.17	18.41	3.24	24.82
Copper pipe to BS2871, Table X brass compression joints and fittings to BS864, two-piece clips at 1250mm maximum centres						
15mm diameter, to timber	m	0.19	1.82	1.77	0.54	4.13
15mm diameter, plugged and screwed	m	0.21	2.02	2.02	0.61	4.65
22mm diameter, to timber	m	0.20	1.92	3.40	0.80	6.12
22mm diameter, plugged and screwed	m	0.22	2.11	3.64	0.86	6.61
28mm diameter, to timber	m	0.22	2.11	5.33	1.12	8.56
28mm diameter, plugged and screwed	m	0.24	2.30	5.58	1.18	9.06
Copper pipe to BS2871, Table Z brass compression joints and fittings to BS864, two-piece clips at 1250mm maximum centres						
15mm diameter, to timber	m	0.19	1.82	1.58	0.51	3.91
15mm diameter, plugged and screwed	m	0.21	2.02	1.82	0.58	4.42
22mm diameter, to timber	m	0.20	1.92	3.10	0.75	5.77
22mm diameter, plugged and screwed	m	0.22	2.11	3.34	0.82	6.27
28mm diameter, to timber	m	0.22	2.11	5.33	1.12	8.56
28mm diameter, plugged and screwed	m	0.24	2.30	5.58	1.18	9.06

Pipework (cont'd)	Unit	Labour hours	Net labour (£)	Net material (£)	O'heads /profit (£)	Total (£)
Copper pipe to BS2871, Table Y dezincification-resistant compression joints and fittings to BS864, laid in trenches						
15mm diameter	m	0.10	0.96	3.21	0.63	4.80
18mm diameter	m	0.11	1.06	4.15	0.78	5.99
22mm diameter	m	0.12	1.15	5.67	1.02	7.84
28mm diameter	m	0.14	1.34	7.39	1.31	10.04
35mm diameter	m	0.16	1.54	13.98	2.33	17.85
42mm diameter	m	0.17	1.63	17.52	2.87	22.02
54mm diameter	m	0.19	1.82	23.90	3.86	29.58
Medium density polyethylene pipes (MDPE) to BS6572 and 6370, brass compression fittings to BS864, laid in trenches						
20mm diameter	m	0.10	0.96	0.69	0.25	1.90
25mm diameter	m	0.10	0.96	0.85	0.27	2.08
32mm diameter	m	0.16	1.54	1.38	0.44	3.36
50mm diameter	m	0.19	1.82	3.47	0.79	6.08
63mm diameter	m	0.20	1.92	5.08	1.05	8.05
Extra over for lead-free pre-soldered capillary joints and fittings to BS864						
8mm fittings						
straight coupling	nr	0.18	1.73	0.67	0.36	2.76
reduced coupling, 8 x 6mm	nr	0.18	1.73	2.06	0.57	4.36
straight female connector, 8mm x 1/4in	nr	0.18	1.73	1.95	0.55	4.23

	Unit	Labour hours	Net labour (£)	Net material (£)	O'heads /profit (£)	Total (£)
straight male connector, 8mm x 1/4in	nr	0. 18	1. 73	1. 37	0. 46	3. 57
reducer, 8 x 6mm	nr	0. 18	1. 73	1. 69	0. 51	3. 93
elbow, 8mm	nr	0. 18	1. 73	1. 60	0. 50	3. 83
street elbow, 8mm	nr	0. 18	1. 73	2. 21	0. 59	4. 53
equal tee, 8mm	nr	0. 24	2. 30	3. 03	0. 80	6. 13
stop end, 8mm	nr	0. 18	1. 73	1. 14	0. 43	3. 30
10mm fittings						
straight coupling	nr	0. 18	1. 73	0. 67	0. 36	2. 76
reduced coupling, 10 x 8mm	nr	0. 18	1. 73	1. 33	0. 46	3. 52
straight male connector, 10mm x 3/8in	nr	0. 18	1. 73	1. 51	0. 49	3. 73
male reducing connector, 10mm x 1/2in	nr	0. 18	1. 73	2. 84	0. 69	5. 26
reducer, 10 x 8mm	nr	0. 18	1. 73	1. 24	0. 45	3. 42
elbow, 10mm	nr	0. 18	1. 73	1. 49	0. 48	3. 70
street elbow, 10mm	nr	0. 18	1. 73	2. 21	0. 59	4. 53
equal tee, 10mm	nr	0. 22	2. 11	2. 84	0. 74	5. 69
stop end, 10mm	nr	0. 18	1. 73	1. 14	0. 43	3. 30
15mm fittings						
made bend	nr	0. 10	0. 96	0. 00	0. 14	1. 10
straight coupling	nr	0. 18	1. 73	0. 19	0. 29	2. 21
reduced coupling, 15 x 8mm	nr	0. 18	1. 73	1. 58	0. 50	3. 81
reduced coupling, 15 x 10mm	nr	0. 18	1. 73	1. 19	0. 44	3. 36
reduced coupling, 15 x 12mm	nr	0. 18	1. 73	1. 27	0. 45	3. 45
adaptor coupling, imperial to metric, 15mm x 1/2in	nr	0. 18	1. 73	0. 92	0. 40	3. 05
straight female connector, 15mm x 1/2in	nr	0. 18	1. 73	1. 37	0. 46	3. 57
female reducing connector, 15mm x 3/8in	nr	0. 18	1. 73	2. 29	0. 60	4. 62
female reducing connector, 15mm x 3/4in	nr	0. 18	1. 73	3. 79	0. 83	6. 35
straight male connector, 15mm x 1/2in	nr	0. 18	1. 73	1. 12	0. 43	3. 28
male reducing connector, 15mm x 3/4in	nr	0. 18	1. 73	2. 70	0. 66	5. 09
male reducing connector, 15mm x 3/8in	nr	0. 18	1. 73	1. 87	0. 54	4. 14

Capillary fittings (cont'd)	Unit	Labour hours	Net labour (£)	Net material (£)	O'heads /profit (£)	Total (£)
lead connector	nr	0.18	1.73	0.89	0.39	3.01
tank connector, 15mm x 1/2in	nr	0.28	2.69	2.84	0.83	6.36
tank connector, 15mm x 3/4in	nr	0.28	2.69	6.36	1.36	10.41
reducer, 15 x 8mm	nr	0.18	1.73	0.96	0.40	3.09
reducer, 15 x 10mm	nr	0.18	1.73	0.96	0.40	3.09
reducer, 15 x 12mm	nr	0.18	1.73	0.95	0.40	3.08
female adaptor, 15mm x 1/2in	nr	0.18	1.73	2.15	0.58	4.46
male adaptor, 15mm x 1/2in	nr	0.18	1.73	2.15	0.58	4.46
adaptor, imperial to metric, 15mm x 1/2in	nr	0.18	1.73	0.84	0.39	2.96
street elbow	nr	0.18	1.73	1.29	0.45	3.47
male elbow, 15mm x 1/2in	nr	0.18	1.73	2.50	0.63	4.86
reducing male elbow, 15mm x 3/8in	nr	0.18	1.73	3.60	0.80	6.13
female elbow, 15mm x 1/2in	nr	0.18	1.73	2.35	0.61	4.69
reducing female elbow, 15mm x 1/4in	nr	0.18	1.73	3.53	0.79	6.05
backplate elbow, 15mm x 1/2in	nr	0.28	2.69	2.76	0.82	6.27
flanged bend, 15mm x 1/2in	nr	0.22	2.11	5.09	1.08	8.28
flanged bend, 15mm x 3/4in	nr	0.22	2.11	5.09	1.08	8.28
overflow bend, 15mm x 1/2in	nr	0.22	2.11	6.24	1.25	9.60
slow bend	nr	0.22	2.11	1.54	0.55	4.20
return bend	nr	0.22	2.11	5.49	1.14	8.74
obtuse elbow	nr	0.18	1.73	1.26	0.45	3.44
tee, reduced branch, largest end, 15mm	nr	0.24	2.30	2.90	0.78	5.98
tee, one end and one branch reduced, largest end, 15mm	nr	0.24	2.30	3.23	0.83	6.36
tee, both ends reduced, largest end, 15mm	nr	0.24	2.30	4.03	0.95	7.28
corner tee	nr	0.24	2.30	5.63	1.19	9.12
sweep tee, 90 degrees	nr	0.24	2.30	3.72	0.90	6.92
offset tee, 15mm equal	nr	0.24	2.30	5.87	1.23	9.40
double sweep tee, 15mm	nr	0.24	2.30	5.11	1.11	8.52
cross equal tee, 15mm	nr	0.28	2.69	4.72	1.11	8.52
stop end, 15mm	nr	0.18	1.73	1.68	0.51	3.92
tap connector, 15mm x 3/4in	nr	0.22	2.11	2.41	0.68	5.20
bent union adaptor, 15mm x 3/4in	nr	0.22	2.11	2.79	0.73	5.6[illegible]
bent male union connector, 15mm x 1/2in	nr	0.22	2.11	5.00	1.07	8.18
bent female union connector, 15mm x 1/2in	nr	0.22	2.11	5.00	1.07	8.18

	Unit	Labour hours	Net labour (£)	Net material (£)	O'heads /profit (£)	Total (£)
22mm fittings						
made bend	nr	0.15	1.44	0.00	0.22	1.66
straight coupling	nr	0.22	2.11	0.44	0.38	2.93
reduced coupling, 22 x 8mm	nr	0.22	2.11	1.87	0.60	4.58
reduced coupling, 22 x 10mm	nr	0.22	2.11	1.87	0.60	4.58
reduced coupling, 22 x 15mm	nr	0.22	2.11	1.07	0.48	3.66
straight female connector, 22mm x 3/4in	nr	0.22	2.11	2.43	0.68	5.22
female reducing connector, 22mm x 1/2in	nr	0.22	2.11	3.62	0.86	6.59
female reducing connector, 22mm x 1in	nr	0.22	2.11	5.00	1.07	8.18
straight male connector, 22mm x 3/4in	nr	0.22	2.11	4.89	1.05	8.05
male reducing connector, 22mm x 1/2in	nr	0.22	2.11	3.41	0.83	6.35
male reducing connector, 22mm x 1in	nr	0.22	2.11	3.37	0.82	6.30
lead connector	nr	0.22	2.11	1.37	0.52	4.00
tank connector, 22mm x 1in	nr	0.30	2.88	10.15	1.95	14.98
reducer, 22 x 12mm	nr	0.22	2.11	1.71	0.57	4.39
female adaptor, 22mm x 3/4in	nr	0.22	2.11	3.21	0.80	6.12
male adaptor, 22mm x 3/4in	nr	0.22	2.11	3.21	0.80	6.12
adaptor, imperial to metric, 22mm x 3/4in	nr	0.22	2.11	0.93	0.46	3.50
reducing elbow, 22 x 15mm	nr	0.22	2.11	4.14	0.94	7.19
street elbow	nr	0.22	2.11	1.98	0.61	4.70
male elbow, 22mm x 3/4in	nr	0.22	2.11	3.56	0.85	6.52
female elbow, 22mm x 3/4in	nr	0.22	2.11	3.23	0.80	6.14
reducing female elbow, 22mm x 1/2in	nr	0.22	2.11	3.43	0.83	6.37
backplate elbow, 22mm x 3/4in	nr	0.24	2.30	6.03	1.25	9.58
flanged bend, 22mm x 3/4in	nr	0.24	2.30	5.18	1.12	8.60
overflow bend, 22mm x 3/4in	nr	0.24	2.30	6.49	1.32	10.11
slow bend	nr	0.24	2.30	2.62	0.74	5.66
return bend	nr	0.24	2.30	6.49	1.32	10.11
obtuse elbow	nr	0.24	2.30	2.07	0.66	5.03
tee, reduced branch, largest end, 22mm	nr	0.28	2.69	3.81	0.97	7.48
tee, both ends reduced, largest end, 22mm	nr	0.28	2.69	4.73	1.11	8.53
corner tee	nr	0.28	2.69	7.25	1.49	11.43
sweep tee, 90 degrees	nr	0.28	2.69	4.64	1.10	8.43

Capillary fittings (cont'd)	Unit	Labour hours	Net labour (£)	Net material (£)	O'heads /profit (£)	Total (£)
sweep tee, reduced branch largest end, 22mm	nr	0.28	2.69	4.89	1.14	8.72
offset tee, 22mm equal	nr	0.28	2.69	6.45	1.37	10.51
double sweep tee, 22mm	nr	0.28	2.69	6.97	1.45	11.11
cross equal tee, 22mm	nr	0.32	3.07	7.57	1.60	12.24
stop end, 22mm	nr	0.20	1.92	2.55	0.67	5.14
tap connector, 22mm x 1/2in	nr	0.24	2.30	4.94	1.09	8.33
bent tap connector, 22mm x 1/2	nr	0.24	2.30	6.01	1.25	9.56
bent tap connector, 22mm x 3/4	nr	0.24	2.30	3.08	0.81	6.19
bent union adaptor, 22mm x 1in	nr	0.24	2.30	4.05	0.95	7.30
bent male union connector, 22mm x 3/4in	nr	0.24	2.30	6.83	1.37	10.50
bent female union connector, 22mm x 3/4in	nr	0.24	2.30	6.83	1.37	10.50
28mm fittings						
made bend	nr	0.20	1.92	0.00	0.29	2.21
straight coupling	nr	0.24	2.30	0.77	0.46	3.53
reduced coupling, 28 x 15mm	nr	0.24	2.30	2.49	0.72	5.51
reduced coupling, 28 x 22mm	nr	0.24	2.30	1.52	0.57	4.39
adaptor coupling, imperial to metric, 1in x 28mm	nr	0.24	2.30	1.69	0.60	4.59
flanged connector	nr	0.24	2.30	22.93	3.78	29.01
straight female connector, 28mm x 1in	nr	0.24	2.30	3.81	0.92	7.03
straight male connector, 28mm x 1in	nr	0.24	2.30	3.23	0.83	6.36
male reducing connector, 28mm x 3/4in	nr	0.24	2.30	4.89	1.08	8.27
lead connector	nr	0.24	2.30	1.97	0.64	4.91
tank connector, 28mm x 1in	nr	0.32	3.07	5.78	1.33	10.18
tank connector, 28mm x 1 1/4in	nr	0.32	3.07	13.86	2.54	19.47
reducer, 28 x 15mm	nr	0.24	2.30	1.35	0.55	4.20
reducer, 28 x 22mm	nr	0.24	2.30	1.14	0.52	3.96
female adaptor, 28mm x 1in	nr	0.24	2.30	4.89	1.08	8.27
male adaptor, 28mm x 1in	nr	0.24	2.30	4.89	1.08	8.27
adaptor, imperial to metric, 28mm x 1in	nr	0.24	2.30	1.60	0.58	4.48
reducing elbow, 28 x 22mm	nr	0.24	2.30	6.62	1.34	10.26
street elbow	nr	0.24	2.30	3.60	0.89	6.79
male elbow, 28mm x 1in	nr	0.24	2.30	5.33	1.14	8.77
female elbow, 28mm x 1in	nr	0.24	2.30	5.05	1.10	8.45
slow bend	nr	0.26	2.50	4.37	1.03	7.90

	Unit	Labour hours	Net labour (£)	Net material (£)	O'heads /profit (£)	Total (£)
return bend	nr	0.26	2.50	9.60	1.81	13.92
obtuse elbow	nr	0.26	2.50	3.62	0.92	7.04
equal tee, 28mm	nr	0.32	3.07	3.39	0.97	7.43
tee, reduced branch, largest end, 28mm	nr	0.32	3.07	3.65	1.01	7.73
tee, one end and one branch reduced, largest end, 28mm	nr	0.32	3.07	4.98	1.21	9.26
tee, both ends reduced, largest end, 28mm	nr	0.32	3.07	4.73	1.17	8.97
sweep tee, 90 degrees	nr	0.32	3.07	9.62	1.90	14.59
sweep tee, reduced branch largest end, 28mm	nr	0.32	3.07	8.28	1.70	13.05
double sweep tee	nr	0.32	3.07	10.59	2.05	15.71
cross equal tee	nr	0.32	3.07	10.85	2.09	16.01
stop end, 28mm	nr	0.22	2.11	2.88	0.75	5.74
bent union adaptor, 28mm x 1 1/4in	nr	0.28	2.69	6.36	1.36	10.41
bent male union connector, 28mm x 1in	nr	0.28	2.69	9.96	1.90	14.55
bent female union connector, 28mm x 1in	nr	0.28	2.69	9.96	1.90	14.55
35mm fittings						
made bend	nr	0.25	2.40	0.00	0.36	2.76
straight coupling	nr	0.28	2.69	1.98	0.70	5.37
reduced coupling, 35 x 22mm	nr	0.28	2.69	3.66	0.95	7.30
reduced coupling, 35 x 28mm	nr	0.28	2.69	4.53	1.08	8.30
adaptor coupling, imperial to metric, 1 1/4in x 35mm	nr	0.28	2.69	3.90	0.99	7.58
flanged connector	nr	0.28	2.69	29.03	4.76	36.48
straight female connector, 35mm x 1 1/4in	nr	0.28	2.69	5.93	1.29	9.91
straight male connector, 35mm x 1 1/4in	nr	0.28	2.69	5.82	1.28	9.79
tank connector, 35mm x 1 1/4in	nr	0.36	3.46	7.90	1.70	13.06
reducer, 35 x 15mm	nr	0.28	2.69	2.72	0.81	6.22
reducer, 35 x 22mm	nr	0.28	2.69	2.60	0.79	6.08
reducer, 35 x 28mm	nr	0.28	2.69	2.60	0.79	6.08
female adaptor, 35mm x 1 1/4in	nr	0.28	2.69	7.21	1.49	11.39
male adaptor, 35mm x 1 1/4in	nr	0.28	2.69	7.21	1.49	11.39
adaptor, imperial to metric, 35mm x 1 1/4in	nr	0.28	2.69	1.98	0.70	5.37
elbow	nr	0.28	2.69	4.23	1.04	7.96

Capillary fittings (cont'd)	Unit	Labour hours	Net labour (£)	Net material (£)	O'heads /profit (£)	Total (£)
street elbow	nr	0.28	2.69	6.69	1.41	10.79
male elbow, 35mm x 1 1/4in	nr	0.28	2.69	7.41	1.51	11.62
female elbow, 35mm x 1 1/4in	nr	0.28	2.69	8.07	1.61	12.37
obtuse elbow	nr	0.28	2.69	8.42	1.67	12.78
equal tee, 35mm	nr	0.36	3.46	7.67	1.67	12.80
tee, reduced branch, largest end, 35mm	nr	0.36	3.46	7.96	1.71	13.13
tee, one end and one branch reduced, largest end, 35mm	nr	0.36	3.46	9.81	1.99	15.26
tee, both ends reduced, largest end, 35mm	nr	0.36	3.46	13.76	2.58	19.80
sweep tee, 90 degrees, 35mm	nr	0.36	3.46	14.47	2.69	20.62
sweep tee, reduced branch, largest end, 35mm	nr	0.36	3.46	14.47	2.69	20.62
stop end, 35mm	nr	0.24	2.30	4.49	1.02	7.81
bent union adaptor, 35mm x 1 1/2in	nr	0.32	3.07	10.28	2.00	15.35
bent male union connector, 35mm x 1 1/4in	nr	0.32	3.07	15.52	2.79	21.38
bent female union connector, 35mm x 1 1/4in	nr	0.32	3.07	15.52	2.79	21.38
42mm fittings						
made bend	nr	0.28	2.69	0.00	0.40	3.09
straight coupling	nr	0.34	3.26	3.00	0.94	7.20
reduced coupling, 42 x 28mm	nr	0.34	3.26	5.78	1.36	10.40
reduced coupling, 42 x 35mm	nr	0.34	3.26	5.78	1.36	10.40
adaptor coupling, imperial to metric, 42mm x 1 1/2in	nr	0.34	3.26	4.95	1.23	9.44
flanged connector	nr	0.34	3.26	34.70	5.69	43.65
straight female connector, 42mm x 1 1/2in	nr	0.34	3.26	7.43	1.60	12.29
straight male connector, 42mm x 1 1/2in	nr	0.34	3.26	7.57	1.62	12.45
tank connector, 42mm x 1 1/2in	nr	0.42	4.03	10.36	2.16	16.55
reducer, 42 x 15mm	nr	0.34	3.26	4.00	1.09	8.35
reducer, 42 x 22mm	nr	0.34	3.26	4.55	1.17	8.98
reducer, 42 x 28mm	nr	0.34	3.26	4.55	1.17	8.98
reducer, 42 x 35mm	nr	0.34	3.26	4.00	1.09	8.35
female adaptor, 42mm x 1 1/2in	nr	0.34	3.26	11.07	2.15	16.48
male adaptor, 42mm x 1 1/2in	nr	0.34	3.26	11.07	2.15	16.48
adaptor, imperial to metric, 1 1/2 x 42mm	nr	0.34	3.26	2.61	0.88	6.75

	Unit	Labour hours	Net labour (£)	Net material (£)	O'heads /profit (£)	Total (£)
elbow	nr	0.34	3.26	6.79	1.51	11.56
street elbow	nr	0.34	3.26	9.90	1.97	15.13
male elbow, 42mm x 1 1/2in	nr	0.32	3.07	8.87	1.79	13.73
obtuse elbow	nr	0.32	3.07	13.11	2.43	18.61
equal tee, 42mm	nr	0.40	3.84	11.52	2.30	17.66
tee, reduced branch, largest end, 42mm	nr	0.40	3.84	14.04	2.68	20.56
tee, one end and one branch reduced, largest end, 42mm	nr	0.40	3.84	14.04	2.68	20.56
tee, both ends reduced, largest end, 42mm	nr	0.40	3.84	22.21	3.91	29.96
sweep tee, 90 degrees, 42mm	nr	0.40	3.84	20.96	3.72	28.52
stop end, 42mm	nr	0.26	2.50	5.96	1.27	9.73
bent male union connector, 42mm x 2in	nr	0.36	3.46	26.81	4.54	34.81
bent female union connector, 42mm x 1 1/2in	nr	0.36	3.46	26.81	4.54	34.81
54mm fittings						
made bend	nr	0.35	3.36	0.00	0.50	3.86
straight coupling	nr	0.38	3.65	7.04	1.60	12.29
reduced coupling, 54 x 35mm	nr	0.38	3.65	8.46	1.82	13.93
reduced coupling, 54 x 42mm	nr	0.38	3.65	8.46	1.82	13.93
flanged connector	nr	0.38	3.65	52.46	8.42	64.53
straight female connector, 54mm x 2in	nr	0.38	3.65	12.69	2.45	18.79
straight male connector, 54mm x 2in	nr	0.38	3.65	11.52	2.28	17.45
tank connector, 54mm x 2in	nr	0.46	4.42	15.81	3.03	23.26
reducer, 54 x 15mm	nr	0.38	3.65	8.20	1.78	13.63
reducer, 54 x 22mm	nr	0.38	3.65	8.20	1.78	13.63
reducer, 54 x 28mm	nr	0.38	3.65	7.13	1.62	12.40
reducer, 54 x 35mm	nr	0.38	3.65	7.13	1.62	12.40
reducer, 54 x 42mm	nr	0.38	3.65	7.13	1.62	12.40
female adaptor, 54mm x 2in	nr	0.38	3.65	13.09	2.51	19.25
male adaptor, 54mm x 2in	nr	0.38	3.65	13.09	2.51	19.25
adaptor, imperial to metric, 54mm x 2in	nr	0.38	3.65	5.34	1.35	10.34
elbow	nr	0.38	3.65	15.46	2.87	21.98
male elbow, 54mm x 2in	nr	0.38	3.65	15.46	2.87	21.98
obtuse elbow	nr	0.38	3.65	17.28	3.14	24.07
equal tee, 54mm	nr	0.46	4.42	21.66	3.91	29.99

Capillary fittings (cont'd)	Unit	Labour hours	Net labour (£)	Net material (£)	O'heads /profit (£)	Total (£)
tee, reduced branch, largest end, 54mm	nr	0.46	4.42	22.21	3.99	30.62
tee, one end and one branch reduced, largest end, 54mm	nr	0.46	4.42	22.21	3.99	30.62
sweep tee, 90 degrees, 54mm	nr	0.46	4.42	26.13	4.58	35.13
stop end, 54mm	nr	0.30	2.88	8.99	1.78	13.65
bent union adaptor, 54mm x 2 1/2in	nr	0.42	4.03	28.55	4.89	37.47
bent male union connector, 54mm x 2in	nr	0.42	4.03	42.72	7.01	53.76
bent female union connector, 54mm x 2in	nr	0.42	4.03	42.72	7.01	53.76
Stopcocks to BS1010						
Gunmetal stopcock with brass headwork, copper x copper						
15mm	nr	0.20	1.92	3.00	0.74	5.66
22mm	nr	0.22	2.11	5.61	1.16	8.88
28mm	nr	0.28	2.69	15.94	2.79	21.42
Dezincification-resistant stopcock, copper x copper						
15mm	nr	0.20	1.92	7.54	1.42	10.88
22mm	nr	0.22	2.11	13.06	2.28	17.45
28mm	nr	0.28	2.69	21.75	3.67	28.11
Gunmetal stopcock, copper x copper						
35mm	nr	0.32	3.07	34.13	5.58	42.78
42mm	nr	0.35	3.36	45.34	7.30	56.01
54mm	nr	0.42	4.03	67.75	10.77	82.55
Gunmetal lockshield stopcock with brass headwork, copper x copper						
15mm	nr	0.20	1.92	7.36	1.39	10.67
22mm	nr	0.22	2.11	10.56	1.90	14.57
28mm	nr	0.28	2.69	18.73	3.21	24.63

	Unit	Labour hours	Net labour (£)	Net material (£)	O'heads /profit (£)	Total (£)
Dezincification-resistant lockshield stopcock, copper x copper						
15mm	nr	0.20	1.92	9.91	1.77	13.60
22mm	nr	0.22	2.11	14.77	2.53	19.41
28mm	nr	0.28	2.69	24.56	4.09	31.34
Gunmetal stopcock, with brass headwork, parallel main thread x copper, 15mm x 1/2in	nr	0.20	1.92	7.90	1.47	11.29
Gunmetal stopcock with brass headwork, parallel main thread with backnut x copper, 15mm x 1/2in	nr	0.20	1.92	8.85	1.62	12.39
Gunmetal stopcock, with brass headwork, lead x copper, 15mm	nr	0.20	1.92	7.06	1.35	10.33
Gunmetal union stopcock with brass headwork, copper x copper union						
15mm	nr	0.20	1.92	8.45	1.56	11.93
22mm	nr	0.22	2.11	12.44	2.18	16.73
28mm	nr	0.28	2.69	20.36	3.46	26.51
Gunmetal union lockshield stopcock with brass headwork, copper x copper union						
15mm	nr	0.20	1.92	9.57	1.72	13.21
22mm	nr	0.22	2.11	14.13	2.44	18.68
28mm	nr	0.28	2.69	23.22	3.89	29.80
Gunmetal double union stopcock with brass headwork, copper x copper union						
15mm	nr	0.20	1.92	9.70	1.74	13.36
22mm	nr	0.22	2.11	13.71	2.37	18.19
28mm	nr	0.28	2.69	24.36	4.06	31.11

Stopcocks (cont'd)	Unit	Labour hours	Net labour (£)	Net material (£)	O'heads /profit (£)	Total (£)
Dezincification-resistant double union stopcock, copper x copper						
15mm	nr	0.20	1.92	15.85	2.67	20.44
22mm	nr	0.22	2.11	19.52	3.24	24.87
28mm	nr	0.28	2.69	36.06	5.81	44.56
Gunmetal double union stopcock, copper x copper						
35mm	nr	0.20	1.92	60.17	9.31	71.40
42mm	nr	0.22	2.11	82.52	12.69	97.32
54mm	nr	0.28	2.69	129.88	19.89	152.46
Gunmetal double union lockshield stopcock, with brass headwork, copper x copper						
15mm	nr	0.20	1.92	10.90	1.92	14.74
22mm	nr	0.22	2.11	22.03	3.62	27.76
28mm	nr	0.28	2.69	27.24	4.49	34.42
Dezincification-resistant double union lockshield stopcock, copper x copper						
15mm	nr	0.20	1.92	11.71	2.04	15.67
22mm	nr	0.22	2.11	21.07	3.48	26.66
28mm	nr	0.28	2.69	38.92	6.24	47.85
Gunmetal stopcock with easy-clean cover and brass headwork, copper x copper						
15mm	nr	0.20	1.92	7.74	1.45	11.11
22mm	nr	0.22	2.11	10.98	1.96	15.05
28mm	nr	0.28	2.69	19.31	3.30	25.30
Gunmetal union stopcock, with easy-clean cover and brass headwork, copper x copper union						
15mm	nr	0.20	1.92	9.43	1.70	13.05

	Unit	Labour hours	Net labour (£)	Net material (£)	O'heads /profit (£)	Total (£)
22mm	nr	0.22	2.11	14.35	2.47	18.93
28mm	nr	0.28	2.69	22.85	3.83	29.37
Gunmetal double union stopcock, with easy-clean cover and brass headwork						
15mm	nr	0.20	1.92	11.30	1.98	15.20
22mm	nr	0.22	2.11	15.60	2.66	20.37
28mm	nr	0.28	2.69	27.44	4.52	34.65
Dezincification-resistant double union stopcock, with easy clean cover, copper x copper						
15mm	nr	0.20	1.92	16.99	2.84	21.75
22mm	nr	0.22	2.11	21.18	3.49	26.78
28mm	nr	0.28	2.69	39.35	6.31	48.35
Gunmetal double union stopcock with easy-clean cover, copper x copper						
35mm	nr	0.35	3.36	65.19	10.28	78.83
42mm	nr	0.42	4.03	90.32	14.15	108.50
54mm	nr	0.46	4.42	145.25	22.45	172.12
Combined gunmetal stopcock and draincock with brass headwork, copper x copper union						
15mm	nr	0.20	1.92	16.70	2.79	21.41
22mm	nr	0.22	2.11	20.55	3.40	26.06
Combined dezincification-resistant stopcock and draincock, copper x copper union, 15mm	nr	0.20	1.92	22.67	3.69	28.28

	Unit	Labour hours	Net labour (£)	Net material (£)	O'heads /profit (£)	Total (£)
Gate valves to BS2871.1						
Gunmetal fullway gate valve, copper x copper						
15mm	nr	0.20	1.92	9.43	1.70	13.05
22mm	nr	0.22	2.11	10.91	1.95	14.97
28mm	nr	0.28	2.69	15.18	2.68	20.55
35mm	nr	0.32	3.07	33.84	5.54	42.45
42mm	nr	0.35	3.36	42.34	6.86	52.56
54mm	nr	0.42	4.03	61.38	9.81	75.22
Gunmetal lockshield gate valve, copper x copper						
15mm	nr	0.20	1.92	9.70	1.74	13.36
22mm	nr	0.22	2.11	11.34	2.02	15.47
28mm	nr	0.28	2.69	17.68	3.06	23.43
35mm	nr	0.32	3.07	34.13	5.58	42.78
42mm	nr	0.35	3.36	42.34	6.86	52.56
54mm	nr	0.42	4.03	61.38	9.81	75.22
Plug cocks						
Nited						
12mm	nr	0.20	1.92	5.98	1.19	9.09
15mm	nr	0.20	1.92	6.98	1.33	10.24
22mm	nr	0.20	1.92	10.49	1.86	14.27
Nited with drop head						
12mm	nr	0.25	2.40	6.60	1.35	10.35
15mm	nr	0.25	2.40	7.92	1.55	11.87
22mm	nr	0.25	2.40	10.84	1.99	15.23
Nited with union outlet						
15mm	nr	0.25	2.40	8.03	1.56	11.99
22mm	nr	0.25	2.40	12.31	2.21	16.92

	Unit	Labour hours	Net labour (£)	Net material (£)	O'heads /profit (£)	Total (£)
Plug stopcock, copper x copper						
15mm	nr	0.20	1.92	8.08	1.50	11.50
22mm	nr	0.20	1.92	12.31	2.13	16.36
Plug stopcock, copper x male thread with backnut						
15mm x 1/2in	nr	0.25	2.40	8.45	1.63	12.48
22mm x 3/4in	nr	0.25	2.40	12.31	2.21	16.92
Manifolds						
Linear flow capillary socket inlet x four soft copper outlets, 22mm x four 8mm	nr	0.60	5.76	7.43	1.98	15.17
Linear flow capillary socket inlet x four soft copper tube outlets, 22mm x four 10mm	nr	0.60	5.76	9.54	2.29	17.60
Linear flow capillary socket inlet x six soft copper tube outlets, 22mm x six 8mm	nr	0.80	7.68	10.63	2.75	21.06
Side entry, capillary socket inlet four compression outlets, 22mm x four 8mm	nr	0.80	7.68	8.64	2.45	18.77
Side entry, capillary socket inlet six compression outlets, 22mm x six 8mm	nr	0.80	7.68	10.30	2.70	20.68
Side entry, capillary socket inlet two compression outlets, 22mm x two 10mm	nr	0.60	5.76	5.64	1.71	13.11
Side entry, capillary socket inlet four compression outlets, 22mm x four 10mm	nr	0.60	5.76	9.04	2.22	17.02

	Unit	Labour hours	Net labour (£)	Net material (£)	O'heads /profit (£)	Total (£)
Valves						
Mistral radiator valve, plain finish, compression inlet x taper male union outlet						
8mm x 1/2in	nr	0.30	2.88	4.91	1.17	8.96
10mm x 1/2in	nr	0.30	2.88	4.91	1.17	8.96
15mm x 1/2in	nr	0.30	2.88	4.91	1.17	8.96
Mistral radiator valve, chromium-plated finish, compression inlet x taper male union outlet						
8mm x 1/2in	nr	0.30	2.88	5.25	1.22	9.35
10mm x 1/2in	nr	0.30	2.88	5.25	1.22	9.35
15mm x 1/2in	nr	0.30	2.88	5.25	1.22	9.35
Maxitwin radiator valve, plain finish, compression connection x taper male union outlet						
two 8mm x 1/2in	nr	0.30	2.88	11.99	2.23	17.10
two 10mm x 1/2in	nr	0.30	2.88	13.43	2.45	18.76
Maxitwin radiator valve, chromium-plated finish, compression connection x taper male union outlet						
two 8mm x 1/2in	nr	0.30	2.88	14.57	2.62	20.07
two 10mm x 1/2in	nr	0.30	2.88	15.99	2.83	21.70
Extra over for dezincification-resistant compression fittings						
15mm						
made bend	nr	0.10	0.96	0.00	0.14	1.10
straight coupling, 15mm	nr	0.16	1.54	1.43	0.45	3.42
male coupling, 15mm x 1/2in	nr	0.16	1.54	1.21	0.41	3.16
male coupling, 15mm x 3/4in	nr	0.16	1.54	2.72	0.64	4.90
male coupling, BSP parallel thread, 15mm x 1/2in	nr	0.16	1.54	1.23	0.42	3.19

	Unit	Labour hours	Net labour (£)	Net material (£)	O'heads /profit (£)	Total (£)
tank coupling with back nut, 15mm x 1/2in	nr	0.16	1.54	3.26	0.72	5.52
female coupling, 15mm x 1/2in	nr	0.16	1.54	1.38	0.44	3.36
female coupling, 15mm x 3/4in	nr	0.16	1.54	2.42	0.59	4.55
elbow, 15mm	nr	0.16	1.54	1.68	0.48	3.70
male elbow, 15mm x 1/2in	nr	0.16	1.54	2.06	0.54	4.14
male elbow, BSP parallel thread, 15mm x 1/2in	nr	0.16	1.54	2.05	0.54	4.13
female elbow, 15mm x 1/2in	nr	0.16	1.54	2.26	0.57	4.37
female elbow, 15mm x 3/4in	nr	0.16	1.54	4.37	0.89	6.80
female wall elbow,15mm x 1/2in	nr	0.16	1.54	3.36	0.73	5.64
tee, 15mm	nr	0.20	1.92	2.32	0.64	4.88
tee with reduced branch, 15 x 15 x 12mm	nr	0.20	1.92	4.10	0.90	6.92
tee,end reduced,15 x 12 x 12mm	nr	0.20	1.92	3.11	0.75	5.78
tee, end and branch reduced, 15 x 12 x 12mm	nr	0.20	1.92	3.53	0.82	6.27
tee, both ends reduced, 15 x 15 x 18mm	nr	0.20	1.92	5.81	1.16	8.89
tee, both ends reduced, 15 x 15 x 22mm	nr	0.20	1.92	4.60	0.98	7.50
male tee, 15mm x 1/2in x 15mm	nr	0.20	1.92	4.57	0.97	7.46
male tee, 15mm x 15mm x 1/2in	nr	0.20	1.92	4.86	1.02	7.80
cross tee, 15mm	nr	0.24	2.30	5.71	1.20	9.21
straight swivel connector, 15mm x 1/2in	nr	0.16	1.54	2.55	0.61	4.70
straight swivel connector, 15mm x 3/4in	nr	0.16	1.54	3.69	0.78	6.01
bent swivel connector, 15mm x 1/2in	nr	0.16	1.54	3.08	0.69	5.31
female tee 15mm x 15mm x 1/2in	nr	0.20	1.92	3.64	0.83	6.39
stop end, 15mm	nr	0.16	1.54	1.68	0.48	3.70
18mm						
made bend	nr	0.15	1.44	0.00	0.22	1.66
straight coupling, 18mm	nr	0.18	1.73	2.64	0.66	5.03
male coupling, 18mm x 1/2in	nr	0.18	1.73	2.23	0.59	4.55
female coupling, 18mm x 1/2in	nr	0.18	1.73	2.36	0.61	4.70
elbow, 18mm	nr	0.18	1.73	2.90	0.69	5.32
male elbow, 18mm x 1/2in	nr	0.18	1.73	2.90	0.69	5.32
female elbow, 18mm x 1/2in	nr	0.18	1.73	3.46	0.78	5.97
female wall elbow,18mm x 1/2in	nr	0.18	1.73	6.54	1.24	9.51
tee, 18mm	nr	0.22	2.11	4.07	0.93	7.11

DZR fittings (cont'd)	Unit	Labour hours	Net labour (£)	Net material (£)	O'heads /profit (£)	Total (£)
tee with reduced branch, 18 x 18 x 12mm	nr	0.22	2.11	5.81	1.19	9.11
tee with reduced branch, 18 x 18 x 15mm	nr	0.22	2.11	4.10	0.93	7.14
tee,end reduced,18 x 12 x 18mm	nr	0.22	2.11	5.48	1.14	8.73
tee,end reduced,18 x 15 x 18mm	nr	0.22	2.11	5.81	1.19	9.11
tee, end and branch reduced, 18 x 12 x 12mm	nr	0.22	2.11	5.48	1.14	8.73
tee, end and branch reduced, 18 x 15 x 12mm	nr	0.22	2.11	5.48	1.14	8.73
tee, end and branch reduced, 18 x 15 x 15mm	nr	0.22	2.11	5.81	1.19	9.11
female adaptor, 18mm x 1/2in	nr	0.18	1.73	2.80	0.68	5.21
reducing set, 18 x12mm	nr	0.18	1.73	0.98	0.41	3.12
reducing set, 18 x 15mm	nr	0.18	1.73	0.98	0.41	3.12
22mm						
made bend	nr	0.15	1.44	0.00	0.22	1.66
straight coupling, 22mm	nr	0.16	1.54	2.22	0.56	4.32
male coupling, 22mm x 1/2in	nr	0.16	1.54	2.41	0.59	4.54
male coupling, 22mm x 3/4in	nr	0.16	1.54	1.79	0.50	3.83
male coupling, 22mm x 1in	nr	0.16	1.54	2.74	0.64	4.92
male coupling, BSP parallel thread, 22mm x 1in	nr	0.16	1.54	2.74	0.64	4.92
tank coupling with back nut, 22mm x 3/4in	nr	0.16	1.54	3.50	0.76	5.80
tank coupling with back nut, 22mm x 1in	nr	0.16	1.54	6.14	1.15	8.83
female coupling, 22mm x 1/2in	nr	0.16	1.54	2.48	0.60	4.62
female coupling, 22mm x 3/4in	nr	0.16	1.54	2.09	0.54	4.17
female coupling, 22mm x 1in	nr	0.16	1.54	3.75	0.79	6.08
elbow, 22mm	nr	0.16	1.54	2.72	0.64	4.90
male elbow, 22mm x 3/4in	nr	0.16	1.54	2.40	0.59	4.53
male elbow, 22mm x 1in	nr	0.16	1.54	4.94	0.97	7.45
male elbow, BSP parallel thread, 22mm x 3/4in	nr	0.16	1.54	2.92	0.67	5.13
male elbow, BSP parallel thread, 22mm x 1in	nr	0.16	1.54	4.94	0.97	7.45
female elbow, 22mm x 3/4in	nr	0.16	1.54	3.26	0.72	5.52
female elbow, 22mm x 1in	nr	0.16	1.54	5.63	1.08	8.25
female wall elbow,22mm x 3/4in	nr	0.16	1.54	6.17	1.16	8.87
tee, 22mm	nr	0.20	1.92	3.91	0.87	6.70

	Unit	Labour hours	Net labour (£)	Net material (£)	O'heads /profit (£)	Total (£)
tee with reduced branch, 22 x 22 x 12mm	nr	0.20	1.92	5.51	1.11	8.54
tee with reduced branch, 22 x 22 x 15mm	nr	0.20	1.92	4.81	1.01	7.74
tee,end reduced,22 x 12 x 22mm	nr	0.20	1.92	5.38	1.09	8.40
tee,end reduced,22 x 15 x 22mm	nr	0.20	1.92	4.81	1.01	7.74
tee, end and branch reduced, 22 x 15 x 15mm	nr	0.20	1.92	4.81	1.01	7.74
tee, both ends reduced, 22 x 22 x 28mm	nr	0.20	1.92	8.39	1.55	11.86
male tee, 22mm x 3/4in	nr	0.20	1.92	7.21	1.37	10.50
male tee, 22mm x 22mm x 3/4in	nr	0.20	1.92	6.55	1.27	9.74
straight swivel connector, 22mm x 3/4in	nr	0.16	1.54	4.33	0.88	6.75
bent swivel connector, 22mm x 3/4in	nr	0.16	1.54	4.99	0.98	7.51
female tee 22mm x 22mm x 1/2in	nr	0.20	1.92	5.02	1.04	7.98
female tee 22mm x 22mm x 3/4in	nr	0.20	1.92	5.11	1.05	8.08
male adaptor, 22mm x 1/2in	nr	0.16	1.54	3.30	0.73	5.57
male adaptor, 22mm x 3/4in	nr	0.16	1.54	3.30	0.73	5.57
female adaptor, 22mm x 1/2in	nr	0.16	1.54	3.30	0.73	5.57
female adaptor, 22mm x 3/4in	nr	0.16	1.54	3.00	0.68	5.22
reducing set, 22 x 12mm	nr	0.16	1.54	1.77	0.50	3.81
reducing set, 22 x 15mm	nr	0.16	1.54	1.33	0.43	3.30
reducing set, 22 x 18mm	nr	0.16	1.54	1.03	0.39	2.96
stop end, 22mm	nr	0.16	1.54	2.52	0.61	4.67
28mm						
made bend	nr	0.20	1.92	0.00	0.29	2.21
straight coupling, 28mm	nr	0.22	2.11	4.27	0.96	7.34
male coupling, 28mm x 3/4in	nr	0.22	2.11	4.15	0.94	7.20
female coupling, 28mm x 1in	nr	0.22	2.11	3.25	0.80	6.16
male coupling, BSP parallel thread 28mm x 1in	nr	0.22	2.11	3.21	0.80	6.12
tank coupling with back nut, 28mm x 1in	nr	0.22	2.11	6.40	1.28	9.79
female coupling, 28mm x 1in	nr	0.22	2.11	3.49	0.84	6.44
elbow, 28mm	nr	0.22	2.11	6.16	1.24	9.51
male elbow, 28mm x 1in	nr	0.22	2.11	4.20	0.95	7.26
male elbow, BSP parallel thread, 28mm x 1in	nr	0.22	2.11	4.20	0.95	7.26
female elbow, 28mm x 1in	nr	0.22	2.11	5.13	1.09	8.33
tee, 28mm	nr	0.26	2.50	7.76	1.54	11.80

DZR fittings (cont'd)	Unit	Labour hours	Net labour (£)	Net material (£)	O'heads /profit (£)	Total (£)
tee with reduced branch, 28 x 28 x 12mm	nr	0.26	2.50	5.23	1.16	8.89
tee with reduced branch, 28 x 28 x 15mm	nr	0.26	2.50	7.94	1.57	12.01
tee with reduced branch, 28 x 28 x 22mm	nr	0.26	2.50	7.94	1.57	12.01
tee,end reduced,28 x 15 x 28mm	nr	0.26	2.50	8.39	1.63	12.52
tee,end reduced,28 x 22 x 28mm	nr	0.26	2.50	7.94	1.57	12.01
tee, end and branch reduced, 28 x 22 x 15mm	nr	0.26	2.50	8.39	1.63	12.52
tee, end and branch reduced, 28 x 22 x 22mm	nr	0.26	2.50	8.39	1.63	12.52
male tee, 28mm x 1in x 28mm	nr	0.26	2.50	12.25	2.21	16.96
female tee 28mm x 28mm x 1/2in	nr	0.26	2.50	7.55	1.51	11.56
female tee, 28mm x 28mm x 1in	nr	0.26	2.50	9.14	1.75	13.39
male adaptor, 28mm x 1in	nr	0.22	2.11	5.41	1.13	8.65
female adaptor, 28mm x 1in	nr	0.22	2.11	5.41	1.13	8.65
reducing set, 28 x 12mm	nr	0.22	2.11	2.13	0.64	4.88
reducing set, 28 x 15mm	nr	0.22	2.11	2.00	0.62	4.73
reducing set, 28 x 22mm	nr	0.22	2.11	2.00	0.62	4.73
stop end, 28mm	nr	0.22	2.11	4.20	0.95	7.26
35mm						
made bend	nr	0.25	2.40	0.00	0.36	2.76
straight coupling, 35mm	nr	0.26	2.50	8.30	1.62	12.42
male coupling, 35mm x 1in	nr	0.26	2.50	8.06	1.58	12.14
male coupling, 35mm x 1 1/4in	nr	0.26	2.50	6.49	1.35	10.34
tank coupling with back nut, 35mm x 1 1/4in	nr	0.26	2.50	7.19	1.45	11.14
female coupling,35mm x 1 1/4in	nr	0.26	2.50	7.12	1.44	11.06
elbow, 35mm	nr	0.26	2.50	10.78	1.99	15.27
male elbow, 35mm x 1 1/4in	nr	0.26	2.50	10.44	1.94	14.88
female elbow, 35mm x 1 1/4in	nr	0.26	2.50	9.95	1.87	14.32
tee, 35mm	nr	0.30	2.88	14.24	2.57	19.69
tee with reduced branch, 35 x 35 x 15mm	nr	0.30	2.88	14.24	2.57	19.69
tee with reduced branch, 35 x 35 x 22mm	nr	0.30	2.88	14.24	2.57	19.69
tee with reduced branch, 35 x 35 x 28mm	nr	0.30	2.88	14.24	2.57	19.69
reducing set, 35 x 15mm	nr	0.26	2.50	3.40	0.89	6.79
reducing set, 35 x 22mm	nr	0.26	2.50	3.40	0.89	6.79
reducing set, 35 x 28mm	nr	0.26	2.50	3.40	0.89	6.79

	Unit	Labour hours	Net labour (£)	Net material (£)	O'heads /profit (£)	Total (£)
tank coupling, 35mm	nr	0.26	2.50	10.61	1.97	15.08
stop end, 35mm	nr	0.26	2.50	6.15	1.30	9.95
42mm						
made bend	nr	0.30	2.88	0.00	0.43	3.31
straight coupling, 42mm	nr	0.28	2.69	10.66	2.00	15.35
male coupling, 42mm x 1 1/2in	nr	0.28	2.69	9.46	1.82	13.97
tank coupling with back nut, 42mm x 1 1/2in	nr	0.28	2.69	9.96	1.90	14.55
female coupling,42mm x 1 1/2in	nr	0.28	2.69	9.67	1.85	14.21
elbow, 42mm	nr	0.28	2.69	15.16	2.68	20.53
male elbow, 42mm x 1 1/2in	nr	0.28	2.69	14.93	2.64	20.26
female elbow, 42mm x 1 1/2in	nr	0.28	2.69	14.93	2.64	20.26
tee, 42mm	nr	0.35	3.36	23.70	4.06	31.12
tee with reduced branch, 42 x 42 x 22mm	nr	0.35	3.36	22.20	3.83	29.39
tee with reduced branch, 42 x 42 x 28mm	nr	0.35	3.36	22.20	3.83	29.39
reducing set, 42 x 15mm	nr	0.28	2.69	6.21	1.33	10.24
reducing set, 42 x 22mm	nr	0.28	2.69	5.37	1.21	9.27
reducing set, 42 x 28mm	nr	0.28	2.69	5.37	1.21	9.27
reducing set, 42 x 35mm	nr	0.28	2.69	5.37	1.21	9.27
stop end, 42mm	nr	0.28	2.69	10.68	2.01	15.38
54mm						
made bend	nr	0.35	3.36	0.00	0.50	3.86
straight coupling, 54mm	nr	0.32	3.07	15.93	2.85	21.85
male coupling, 54mm x 2in	nr	0.32	3.07	13.62	2.50	19.19
tank coupling with back nut, 54mm x 2in	nr	0.32	3.07	15.34	2.76	21.17
female coupling, 54mm x 2in	nr	0.32	3.07	14.00	2.56	19.63
elbow, 54mm	nr	0.32	3.07	21.30	3.66	28.03
male elbow, 54mm x 2in	nr	0.32	3.07	22.16	3.78	29.01
tee, 54mm	nr	0.40	3.84	37.14	6.15	47.13
tee with reduced branch, 54 x 54 x 28mm	nr	0.40	3.84	37.14	6.15	47.13
reducing set, 54 x 22mm	nr	0.32	3.07	10.31	2.01	15.39
reducing set, 54 x 28mm	nr	0.32	3.07	10.31	2.01	15.39
reducing set, 54 x 35mm	nr	0.32	3.07	8.97	1.81	13.85
reducing set, 54 x 42mm	nr	0.32	3.07	8.97	1.81	13.85

	Unit	Labour hours	Net labour (£)	Net material (£)	O'heads /profit (£)	Total (£)
Extra over for brass compression fittings						
15mm						
made bend	nr	0.10	0.96	0.00	0.14	1.10
straight coupling	nr	0.16	1.54	0.96	0.38	2.88
male coupling, extended thread 15mm x 1/2in	nr	0.20	1.92	3.26	0.78	5.96
lead coupling	nr	0.24	2.30	2.43	0.71	5.44
bent union radiator plain, 15mm x 1/2in	nr	0.16	1.54	3.88	0.81	6.23
air release elbow, 15mm	nr	0.16	1.54	5.15	1.00	7.69
slow bend	nr	0.16	1.54	5.19	1.01	7.74
male elbow, 15mm x 1/2in	nr	0.16	1.54	1.46	0.45	3.45
male elbow, 15mm x 3/4in	nr	0.16	1.54	4.31	0.88	6.73
male elbow, BSP parallel thread, 15mm x 1/2in	nr	0.16	1.54	1.50	0.46	3.50
female elbow, 15mm x 1/2in	nr	0.16	1.54	2.01	0.53	4.08
female wall elbow,15mm x 1/2in	nr	0.16	1.54	2.97	0.68	5.19
female wall elbow,15mm x 3/4in	nr	0.16	1.54	6.42	1.19	9.15
female tee with back plate, 15 x 15 x 15mm	nr	0.24	2.30	5.26	1.13	8.69
offset tee, 15 x 15 x 15mm	nr	0.20	1.92	8.84	1.61	12.37
male adaptor, 15mm x 1/2in	nr	0.16	1.54	1.66	0.48	3.68
reducing set, 15 x 8mm	nr	0.16	1.54	1.17	0.41	3.12
reducing set, 15 x 10mm	nr	0.16	1.54	1.17	0.41	3.12
reducing set, 15 x 12mm	nr	0.16	1.54	1.18	0.41	3.13
sweep tee, 15mm	nr	0.20	1.92	8.82	1.61	12.35
corner tee, 15 x 15 x 15mm	nr	0.20	1.92	9.04	1.64	12.60
22mm						
made bend	nr	0.15	1.44	0.00	0.22	1.66
straight coupling	nr	0.16	1.54	1.62	0.47	3.63
male coupling, extended thread, 22mm x 3/4in	nr	0.16	1.54	4.14	0.85	6.53
female coupling, 22mm x 1in	nr	0.16	1.54	2.64	0.63	4.81
lead coupling, 22mm	nr	0.16	1.54	3.43	0.75	5.72
air release elbow, 22mm	nr	0.16	1.54	6.50	1.21	9.25
slow bend, 22mm	nr	0.16	1.54	8.51	1.51	11.56
male elbow, 22mm x 3/4in	nr	0.16	1.54	1.87	0.51	3.92
male elbow, 22mm x 1in	nr	0.16	1.54	3.31	0.73	5.58

	Unit	Labour hours	Net labour (£)	Net material (£)	O'heads /profit (£)	Total (£)
male elbow, BSP parallel thread, 22mm x 3/4in	nr	0.16	1.54	1.93	0.52	3.99
male elbow, BSP parallel thread, 22mm x 1in	nr	0.16	1.54	3.31	0.73	5.58
female elbow, 22mm x 3/4in	nr	0.16	1.54	2.92	0.67	5.13
female elbow, 22mm x 1in	nr	0.16	1.54	4.88	0.96	7.38
bent cylinder connectors, 22mm x 1in	nr	0.16	1.54	7.33	1.33	10.20
tee with reduced branch, 22 x 22 x 15mm	nr	0.20	1.92	4.52	0.97	7.41
tee,end reduced,22 x 15 x 22mm	nr	0.20	1.92	4.61	0.98	7.51
tee, end and branch reduced, 22 x 15 x 15mm	nr	0.20	1.92	4.63	0.98	7.53
straight swivel connector, 22mm x 1/2in	nr	0.16	1.54	3.88	0.81	6.23
bent swivel connector, 22mm x 1/2in	nr	0.16	1.54	5.49	1.05	8.08
offset tee, 22 x 22 x 15mm	nr	0.16	1.54	13.02	2.18	16.74
tank coupling, extended thread, 22mm	nr	0.16	1.54	5.43	1.05	8.02
sweep tee, 22mm	nr	0.16	1.54	9.55	1.66	12.75
28mm						
made bend	nr	0.20	1.92	0.00	0.29	2.21
straight coupling	nr	0.22	2.11	3.96	0.91	6.98
male coupling, 28mm x 1 1/4in	nr	0.22	2.11	7.89	1.50	11.50
female coupling, 28mm x 3/4in	nr	0.22	2.11	4.31	0.96	7.38
female coupling, 28mm x 1in	nr	0.22	2.11	2.74	0.73	5.58
slow bend, 28mm	nr	0.22	2.11	11.45	2.03	15.59
male elbow, 28mm x 1in	nr	0.22	2.11	4.01	0.92	7.04
male elbow, BSP parallel thread, 28mm x 1in	nr	0.22	2.11	4.01	0.92	7.04
female elbow, 28mm x 1in	nr	0.22	2.11	4.86	1.05	8.02
tee, 28mm	nr	0.26	2.50	7.43	1.49	11.42
female tee,28mm x 28mm x 3/4in	nr	0.26	2.50	9.52	1.80	13.82
tank coupling, 28mm	nr	0.22	2.11	4.95	1.06	8.12

Stopcocks to BS1010

Gunmetal stopcocks with brass headwork, copper x copper

	Unit	Labour hours	Net labour (£)	Net material (£)	O'heads /profit (£)	Total (£)
15mm	nr	0.19	1.82	4.07	0.88	6.77

Stopcocks (cont'd)	Unit	Labour hours	Net labour (£)	Net material (£)	O'heads /profit (£)	Total (£)
22mm	nr	0.21	2.02	6.96	1.35	10.33
28mm	nr	0.27	2.59	18.65	3.19	24.43
Gunmetal lockshield stopcock with brass headwork, copper x copper						
15mm	nr	0.19	1.82	8.89	1.61	12.32
22mm	nr	0.21	2.02	12.46	2.17	16.65
28mm	nr	0.27	2.59	24.04	3.99	30.62
Stopcock, dezincification-resistant, copper x copper						
15mm	nr	0.19	1.82	9.50	1.70	13.02
22mm	nr	0.21	2.02	15.59	2.64	20.25
28mm	nr	0.27	2.59	25.78	4.26	32.63
35mm	nr	0.31	2.98	47.44	7.56	57.98
42mm	nr	0.34	3.26	67.87	10.67	81.80
54mm	nr	0.41	3.94	92.44	14.46	110.84
Lockshield stopcock, dezincification-resistant, copper x copper						
15mm	nr	0.19	1.82	10.92	1.91	14.65
22mm	nr	0.21	2.02	17.11	2.87	22.00
Gunmetal stopcock with brass headwork, parallel male thread x copper						
15mm x 1/2in	nr	0.19	1.82	8.07	1.48	11.37
22mm x 3/4in	nr	0.21	2.02	11.21	1.98	15.21
Gunmetal stopcock with brass headwork, copper x lead inlet, 15mm	nr	0.19	1.82	8.09	1.49	11.40
Gunmetal stopcock with brass headwork, and easy-clean cover, copper x copper						
15mm	nr	0.19	1.82	9.77	1.74	13.33
22mm	nr	0.21	2.02	13.79	2.37	18.18
28mm	nr	0.27	2.59	24.69	4.09	31.37

	Unit	Labour hours	Net labour (£)	Net material (£)	O'heads /profit (£)	Total (£)
Combined gunmetal stopcock and draincock, with brass headwork, copper x copper						
15mm	nr	0.19	1.82	13.67	2.32	17.81
22mm	nr	0.21	2.02	21.05	3.46	26.53
Combined stopcock and draincock, dezincification-resistant, copper x copper						
15mm	nr	0.19	1.82	17.62	2.92	22.36
22mm	nr	0.21	2.02	25.82	4.18	32.02
Gunmetal stopcock, soft copper x soft copper						
15mm	nr	0.19	1.82	15.04	2.53	19.39
28mm	nr	0.27	2.59	28.19	4.62	35.40
Gunmetal stopcock with brass headwork, polyethylene x polyethylene						
20mm	nr	0.19	1.82	11.02	1.93	14.77
25mm	nr	0.21	2.02	17.62	2.95	22.59
Stopcock, dezincification-resistant, polyethylene x polyethylene						
20mm	nr	0.19	1.82	14.79	2.49	19.10
25mm	nr	0.21	2.02	23.81	3.87	29.70
32mm	nr	0.27	2.59	32.04	5.19	39.82
50mm	nr	0.31	2.98	80.36	12.50	95.84
63mm	nr	0.41	3.94	113.34	17.59	134.87
Gunmetal stopcock for underground use, dezincification-resistant, polyethylene x polyethylene						
20mm	nr	0.19	1.82	13.04	2.23	17.09
25mm	nr	0.21	2.02	20.96	3.45	26.43

Stopcocks (cont'd)	Unit	Labour hours	Net labour (£)	Net material (£)	O'heads /profit (£)	Total (£)
Gunmetal stopcock with brass headwork, polyethylene x copper						
20 x 15mm	nr	0.19	1.82	10.21	1.80	13.83
25 x 15mm	nr	0.21	2.02	17.71	2.96	22.69
25 x 22mm	nr	0.29	2.78	18.90	3.25	24.93
Combined gunmetal stopcock and draincock with brass headwork, polyethylene x polyethylene, 20mm	nr	0.19	1.82	19.78	3.24	24.84
Combined gunmetal stopcock and draincock with brass headwork, polyethylene x copper						
20 x 15mm	nr	0.19	1.82	19.30	3.17	24.29
Combined gunmetal stopcock and draincock, dezincification-resistant, polyethylene x copper						
20 x 15mm	nr	0.19	1.82	28.52	4.55	34.89
25 x 15mm	nr	0.21	2.02	29.73	4.76	36.51
25 x 22mm	nr	0.23	2.21	34.22	5.46	41.89
Gunmetal stopcock with brass headwork, polyethylene x polyethylene						
3/8in	nr	0.19	1.82	10.84	1.90	14.56
1/2in	nr	0.21	2.02	11.12	1.97	15.11
3/4in	nr	0.23	2.21	16.78	2.85	21.84
1in	nr	0.27	2.59	24.25	4.03	30.87
Gate valves to BS2871.1						
Brass gate valve, copper x copper						
15mm	nr	0.19	1.82	6.64	1.27	9.73
22mm	nr	0.21	2.02	7.73	1.46	11.21
28mm	nr	0.27	2.59	13.48	2.41	18.48

	Unit	Labour hours	Net labour (£)	Net material (£)	O'heads /profit (£)	Total (£)
Brass lockshield gate valve, copper x copper						
15mm	nr	0.19	1.82	6.69	1.28	9.79
22mm	nr	0.21	2.02	7.89	1.49	11.40
28mm	nr	0.27	2.59	14.30	2.53	19.42
Gunmetal fullway gate valve, copper x copper						
15mm	nr	0.19	1.82	11.50	2.00	15.32
22mm	nr	0.21	2.02	13.59	2.34	17.95
28mm	nr	0.27	2.59	18.47	3.16	24.22
35mm	nr	0.36	3.46	28.28	4.76	36.50
42mm	nr	0.50	4.80	37.16	6.29	48.25
54mm	nr	0.60	5.76	51.29	8.56	65.61
Gunmetal lockshield gate valve, copper x copper						
15mm	nr	0.19	1.82	11.85	2.05	15.72
22mm	nr	0.21	2.02	14.09	2.42	18.53
28mm	nr	0.27	2.59	21.56	3.62	27.77
35mm	nr	0.36	3.46	27.58	4.66	35.70
42mm	nr	0.50	4.80	37.81	6.39	49.00
54mm	nr	0.60	5.76	53.74	8.92	68.43
Check valves						
Dezincification-resistant alloy check valve with compression nuts						
15mm	nr	0.19	1.82	12.79	2.19	16.80
22mm	nr	0.21	2.02	13.67	2.35	18.04
28mm	nr	0.27	2.59	21.56	3.62	27.77
Dezincification-resistant alloy check valve with female BSP parallel threads						
1/2in	nr	0.19	1.82	12.40	2.13	16.35
3/4in	nr	0.21	2.02	13.79	2.37	18.18
1in	nr	0.27	2.59	19.69	3.34	25.62

	Unit	Labour hours	Net labour (£)	Net material (£)	O'heads /profit (£)	Total (£)
Drainage fittings to BS2879 Type 1						
Dezincification-resistant stopcock adaptor, copper x copper, 15mm	nr	0.20	1.92	6.63	1.28	9.83
Draw-off coupling, copper x copper, 15mm	nr	0.20	1.92	6.05	1.20	9.17
Dezincification-resistant draw-off coupling, copper x copper						
15mm	nr	0.20	1.92	8.13	1.51	11.56
22mm	nr	0.25	2.40	9.43	1.77	13.60
Dezincification-resistant male draw-off elbow, copper x parallel main thread						
15mm x 1/2in	nr	0.20	1.92	6.22	1.22	9.36
22mm x 3/4in	nr	0.25	2.40	6.99	1.41	10.80
28mm x 1in	nr	0.30	2.88	11.41	2.14	16.43
Brass draw-off elbow, copper to copper						
15mm	nr	0.20	1.92	6.04	1.19	9.15
22mm	nr	0.25	2.40	6.92	1.40	10.72
28mm	nr	0.30	2.88	10.67	2.03	15.58
Dezincification-resistant draw-off elbow, copper x copper						
15mm	nr	0.20	1.92	8.94	1.63	12.49
22mm	nr	0.25	2.40	11.03	2.01	15.44
Brass draincock, parallel thread connector, 1/2in	nr	0.20	1.92	1.85	0.57	4.34
Dezincification-resistant drain-cock, parallel threaded connector, 1/2in	nr	0.20	1.92	4.24	0.92	7.08

	Unit	Labour hours	Net labour (£)	Net material (£)	O'heads /profit (£)	Total (£)
Servicing valves						
Straight pattern, copper x copper						
15 x 15mm	nr	0.20	1.92	3.70	0.84	6.46
22 x 22mm	nr	0.25	2.40	6.04	1.27	9.71
Straight pattern, copper x swivel end, 15mm x 1/2in	nr	0.20	1.92	3.97	0.88	6.77
Bent pattern, copper x swivel end, 15mm x 1/2in	nr	0.20	1.92	3.96	0.88	6.76
Appliance valves						
Chromium-plated stopvalve, 15mm x 3/4in	nr	0.20	1.92	12.63	2.18	16.73
Brass stopvalve, 15mm x 3/4in	nr	0.20	1.92	6.05	1.20	9.17
Brass in-line valve, 15mm x 3/4in	nr	0.20	1.92	2.97	0.73	5.62
Brass elbow valve, 15mm x 3/4in	nr	0.20	1.92	2.97	0.73	5.62
Brass tee valve, 15mmx15mmx3/4in	nr	0.20	1.92	4.87	1.02	7.81
Cold water storage tanks						
Galvanized steel cistern, BS417 Part A with cover, reference						
SC10, 18 litres	nr	0.60	5.76	35.16	6.14	47.06
SC15, 36 litres	nr	0.65	6.24	39.83	6.91	52.98
SC20, 54 litres	nr	0.65	6.24	45.91	7.82	59.97
SC25, 68 litres	nr	0.70	6.72	47.66	8.16	62.54
SC30, 86 litres	nr	0.75	7.20	49.43	8.49	65.12
SC40, 114 litres	nr	0.80	7.68	56.23	9.59	73.50
SC50, 154 litres	nr	0.85	8.16	79.96	13.22	101.34
SC60, 191 litres	nr	0.85	8.16	84.91	13.96	107.03
SC70, 227 litres	nr	1.00	9.60	92.89	15.37	117.86
SC80, 260 litres	nr	1.10	10.56	97.62	16.23	124.41
SC100/2, 330 litres	nr	1.30	12.48	107.31	17.97	137.76
SC100/1, 320 litres	nr	1.30	12.48	108.92	18.21	139.61
SC125, 414 litres	nr	1.50	14.40	168.16	27.38	209.94

Cold water storage tanks (cont'd)	Unit	Labour hours	Net labour (£)	Net material (£)	O'heads /profit (£)	Total (£)
Cutting holes for connectors						
13mm	nr	0.10	0.96	0.00	0.14	1.10
18mm	nr	0.12	1.15	0.00	0.17	1.32
25mm	nr	0.15	1.44	0.00	0.22	1.66
32mm	nr	0.20	1.92	0.00	0.29	2.21
Plastic cistern, BS4213 with lid, reference						
PC4, 18 litres	nr	0.60	5.76	6.71	1.87	14.34
PC15, 68 litres	nr	0.70	6.72	23.59	4.55	34.86
PC20, 91 litres	nr	0.75	0.00	25.02	3.75	28.77
PC25, 114 litres	nr	0.80	7.68	31.88	5.93	45.49
PC40, 182 litres	nr	0.90	8.64	57.28	9.89	75.81
PC50, 227 litres	nr	1.00	9.60	46.45	8.41	64.46
Cutting holes for connectors						
13mm	nr	0.10	0.96	0.00	0.14	1.10
18mm	nr	0.12	1.15	0.00	0.17	1.32
25mm	nr	0.15	1.44	0.00	0.22	1.66
32mm	nr	0.20	1.92	0.00	0.29	2.21
Hot water tanks						
Galvanized steel tank, BS417 with cover, reference						
T25/1, 36 litres	nr	0.65	6.24	114.48	18.11	138.83
T25/2, 36 litres	nr	0.65	6.24	116.27	18.38	140.89
T30/1, 86 litres	nr	0.75	7.20	122.66	19.48	149.34
T30/2, 86 litres	nr	0.75	7.20	124.17	19.71	151.08
T40, 114 litres	nr	0.80	7.68	150.39	23.71	181.78
Hot water copper cylinders						
Direct cylinders to BS 699 grade 3 insulated, reference						
ref 5, 98 litres	nr	0.50	4.80	94.10	14.84	113.74
ref 7, 120 litres	nr	0.55	5.28	105.03	16.55	126.86
ref 8, 148 litres	nr	0.65	6.24	114.64	18.13	139.01
ref 9, 166 litres	nr	0.85	8.16	134.73	21.43	164.32

	Unit	Labour hours	Net labour (£)	Net material (£)	O'heads /profit (£)	Total (£)
Indirect cylinder to BS699 grade 3 insulated, reference						
ref 2, 96 litres	nr	0.50	4.80	115.96	18.11	138.87
ref 3, 114 litres	nr	0.55	5.28	123.49	19.32	148.09
ref 7, 117 litres	nr	0.55	5.28	121.87	19.07	146.22
ref 8, 140 litres	nr	0.65	6.24	137.97	21.63	165.84
ref 9, 162 litres	nr	0.70	6.72	176.23	27.44	210.39
Fortic single-feed cylinder to BS3198, 4 connections, plain						
900 x 400mm, capacity 65 litres hot, 20 litres cold						
F1, direct	nr	2.80	26.88	143.44	25.55	195.87
F3, indirect coil	nr	2.80	26.88	179.18	30.91	236.97
F6, primatic indirect	nr	2.80	26.88	188.64	32.33	247.85
F8, primatic indirect	nr	2.80	26.88	201.20	34.21	262.29
900 x 450mm, capacity 85 litres hot, 25 litres cold						
F1, direct	nr	3.00	28.80	147.71	26.48	202.99
F3, indirect coil	nr	3.00	28.80	185.25	32.11	246.16
F6, primatic indirect	nr	3.00	28.80	194.25	33.46	256.51
F8, primatic indirect	nr	3.00	28.80	205.34	35.12	269.26
1075 x 450mm, capacity 115 litres hot, 25 litres cold						
F1, direct	nr	3.20	30.72	162.49	28.98	222.19
F3, indirect coil	nr	3.20	30.72	201.48	34.83	267.03
F6, primatic indirect	nr	3.20	30.72	211.97	36.40	279.09
F8, primatic indirect	nr	3.20	30.72	223.50	38.13	292.35
1200 x 450mm, capacity 115 litres hot, 45 litres cold						
F1, direct	nr	3.40	32.64	172.99	30.84	236.47
F3, indirect coil	nr	3.40	32.64	211.83	36.67	281.14
F6, primatic indirect	nr	3.40	32.64	221.88	38.18	292.70
F8, primatic indirect	nr	3.40	32.64	234.29	40.04	306.97

Cylinders (cont'd)	Unit	Labour hours	Net labour (£)	Net material (£)	O'heads /profit (£)	Total (£)
1300 x 450mm, capacity 130 litres hot, 45 litres cold						
F1, direct	nr	3.60	34.56	201.94	35.48	271.98
F3, indirect coil	nr	3.60	34.56	236.80	40.70	312.06
F6, primatic indirect	nr	3.60	34.56	260.43	44.25	339.24
F8, primatic indirect	nr	3.60	34.56	273.28	46.18	354.02
1200 x 500mm, capacity 150 litres hot, 45 litres cold						
F1, direct	nr	3.80	36.48	245.66	42.32	324.46
F3, indirect coil	nr	3.80	36.48	281.12	47.64	365.24
F6, primatic indirect	nr	3.80	36.48	298.54	50.25	385.27
F8, primatic indirect	nr	3.80	36.48	310.37	52.03	398.88
1400 x 500mm, capacity 115 litres hot, 115 litres cold						
F1, direct	nr	4.00	38.40	262.21	45.09	345.70
F3, indirect coil	nr	4.00	38.40	291.75	49.52	379.67
F6, primatic indirect	nr	4.00	38.40	313.61	52.80	404.81
F8, primatic indirect	nr	4.00	38.40	324.69	54.46	417.55
Insulation						
Preformed pipe lagging, fire-retardant foam, 22mm thick to pipe size						
15mm	m	0.08	0.77	0.75	0.23	1.75
22mm	m	0.08	0.77	0.92	0.25	1.94
28mm	m	0.10	0.96	1.10	0.31	2.37
50mm thick glass fibre filled insulating jacket, fixing bands to tank size						
445 x 305 x 300mm	nr	0.35	3.36	3.12	0.97	7.45
495 x 368 x 362mm	nr	0.40	3.84	5.68	1.43	10.95
630 x 450 x 420mm	nr	0.45	4.32	6.95	1.69	12.96
665 x 490 x 515mm	nr	0.50	4.80	7.61	1.86	14.27
700 x 540 x 535mm	nr	0.60	5.76	10.75	2.48	18.99
995 x 605 x 595mm	nr	0.70	6.72	13.35	3.01	23.08

T

HEATING SYSTEMS

	Unit	Material supply (£)	Material waste (%)	Total (£)
T10 GAS/OIL FIRED BOILERS				
MATERIAL COSTS				
Gas fired free-standing boiler for domestic central heating and indirect hot water				
Balanced flue, output				
40,000 Btu	nr	562.30	1.0	567.92
50,000 Btu	nr	599.60	1.0	605.60
60,000 Btu	nr	642.20	1.0	648.62
80,000 Btu	nr	879.90	1.0	888.70
100,000 Btu	nr	1102.20	1.0	1113.22
Open flue, output				
40,000 Btu	nr	446.70	1.0	451.17
50,000 Btu	nr	470.80	1.0	475.51
60,000 Btu	nr	498.40	1.0	503.38
80,000 Btu	nr	638.30	1.0	644.68
100,000 Btu	nr	829.70	1.0	838.00
Gas fired wall-hung boiler for domestic central heating and indirect heating and indirect hot water				
Balanced flue, output				
30,000 Btu	nr	429.60	1.0	433.90
40,000 Btu	nr	485.40	1.0	490.25
50,000 Btu	nr	562.90	1.0	568.53
60,000 Btu	nr	684.50	1.0	691.35
80,000 Btu	nr	1126.90	1.0	1138.17
Open flue, output				
35,000 Btu	nr	414.20	1.0	418.34
50,000 Btu	nr	458.80	1.0	463.39

	Unit	Material supply (£)	Material waste (%)	Total (£)
Oil fired boiler for domestic central heating and indirect heating and indirect hot water				
Balanced flue, output				
50,000 Btu	nr	918.00	1.0	927.18
60,000 Btu	nr	951.00	1.0	960.51
80,000 Btu	nr	1067.00	1.0	1077.67
100,000 Btu	nr	1194.00	1.0	1205.94
Conventional flue output				
50,000 Btu	nr	932.00	1.0	941.32
60,000 Btu	nr	962.00	1.0	971.62
80,000 Btu	nr	1075.00	1.0	1085.75
100,000 Btu	nr	1184.00	1.0	1195.84
Fibre cement light quality flue pipe and fittings				
75mm	m	8.56	5.0	8.99
bend	nr	5.69	2.5	5.83
bend with door	nr	9.95	2.5	10.20
equal tee	nr	7.13	2.5	7.31
cone caps	nr	12.26	2.5	12.57
GC1 terminal	nr	21.73	2.5	22.27
GLC terminal	nr	16.97	2.5	17.39
H piece	nr	27.35	2.5	28.03
blank cap	nr	3.29	2.5	3.37
100mm	m	11.49	5.0	12.06
bend	nr	7.10	2.5	7.28
bend with door	nr	11.61	2.5	11.90
equal tee	nr	8.84	2.5	9.06
cone caps	nr	16.03	2.5	16.43
GC1 terminal	nr	23.07	2.5	23.65
GLC terminal	nr	21.91	2.5	22.46
H piece	nr	27.35	2.5	28.03
blank cap	nr	4.14	2.5	4.24

	Unit	Material supply (£)	Material waste (%)	Total (£)
125mm	m	14.54	5.0	15.27
bend	nr	9.12	2.5	9.35
bend with door	nr	14.70	2.5	15.07
equal tee	nr	11.57	2.5	11.86
cone caps	nr	18.13	2.5	18.58
GC1 terminal	nr	26.17	2.5	26.82
GLC terminal	nr	23.56	2.5	24.15
H piece	nr	33.78	2.5	34.62
blank cap	nr	5.10	2.5	5.23
150mm	m	18.60	5.0	19.53
bend	nr	10.94	2.5	11.21
bend with door	nr	16.40	2.5	16.81
equal tee	nr	13.58	2.5	13.92
cone caps	nr	21.75	2.5	22.29
GC1 terminal	nr	34.80	2.5	35.67
GLC terminal	nr	25.90	2.5	26.55
H piece	nr	40.96	2.5	41.98
blank cap	nr	7.33	2.5	7.51
Fibre cement heavy quality flue pipe and fittings				
100mm	m	14.75	5.0	15.49
bend	nr	10.08	2.5	10.33
bend with door	nr	16.73	2.5	17.15
equal tee	nr	10.42	2.5	10.68
H piece	nr	29.01	2.5	29.74
blank cap	nr	4.47	2.5	4.58
125mm	m	19.65	5.0	20.63
bend	nr	12.48	2.5	12.79
bend with door	nr	20.61	2.5	21.13
equal tee	nr	13.04	2.5	13.37
H piece	nr	36.27	2.5	37.18
blank cap	nr	6.14	2.5	6.29

Flue pipes (cont'd)	Unit	Material supply (£)	Material waste (%)	Total (£)
150mm	m	25.49	5.0	26.76
bend	nr	14.37	2.5	14.73
bend with door	nr	24.22	2.5	24.83
equal tee	nr	15.26	2.5	15.64
H piece	nr	43.50	2.5	44.59
blank cap	nr	7.84	2.5	8.04
175mm	m	34.79	5.0	36.53
bend	nr	22.88	2.5	23.45
bend with door	nr	33.53	2.5	34.37
equal tee	nr	23.41	2.5	24.00
GLC terminal	nr	35.97	2.5	36.87
H piece	nr	74.46	2.5	76.32
blank cap	nr	11.22	2.5	11.50
200mm	m	37.84	5.0	39.73
bend	nr	35.99	2.5	36.89
bend with door	nr	46.89	2.5	48.06
equal tee	nr	37.20	2.5	38.13
GLC terminal	nr	37.37	2.5	38.30
H piece	nr	114.92	2.5	117.79

	Unit	Labour hours	Net labour (£)	Net material (£)	O'heads /profit (£)	Total (£)
T10 GAS/OIL FIRED BOILERS						
Gas fired free-standing boiler for domestic central heating and indirect hot water						
Balanced flue, output						
40,000 Btu	nr	4.80	46.08	579.34	93.81	719.23
50,000 Btu	nr	4.80	46.08	617.78	99.58	763.44
60,000 Btu	nr	4.80	46.08	661.66	106.16	813.90
80,000 Btu	nr	4.80	46.08	888.39	140.17	1074.64
100,000 Btu	nr	4.80	46.08	1135.59	177.25	1358.92
Open flue, output						
40,000 Btu	nr	4.20	40.32	460.24	75.08	575.64
50,000 Btu	nr	4.20	40.32	485.07	78.81	604.20
60,000 Btu	nr	4.20	40.32	513.49	83.07	636.88
80,000 Btu	nr	4.20	40.32	657.64	104.69	802.65
100,000 Btu	nr	4.20	40.32	854.84	134.27	1029.43
Gas fired wall-hung boiler for domestic central heating and indirect heating and indirect hot water						
Balanced flue, output						
30,000 Btu	nr	4.80	46.08	442.62	73.30	562.01
40,000 Btu	nr	4.80	46.08	500.10	81.93	628.11
50,000 Btu	nr	4.80	46.08	579.96	93.91	719.95
60,000 Btu	nr	4.80	46.08	705.24	112.70	864.02
80,000 Btu	nr	4.80	46.08	1161.05	181.07	1388.20
Open flue, output						
35,000 Btu	nr	4.20	40.32	426.75	70.06	537.13
50,000 Btu	nr	4.20	40.32	472.70	76.95	589.97

	Unit	Labour hours	Net labour (£)	Net material (£)	O'heads /profit (£)	Total (£)
Oil fired boiler for domestic central heating and indirect heating and indirect hot water						
Balanced flue, output						
50,000 Btu	nr	4.80	46.08	945.92	148.80	1140.80
60,000 Btu	nr	4.80	46.08	979.82	153.89	1179.79
80,000 Btu	nr	4.80	46.08	1099.33	171.81	1317.22
100,000 Btu	nr	4.80	46.08	1230.18	191.44	1467.70
Conventional flue output						
50,000 Btu	nr	4.20	40.32	960.26	150.09	1150.67
60,000 Btu	nr	4.20	40.32	991.15	154.72	1186.19
80,000 Btu	nr	4.20	40.32	1107.58	172.19	1320.09
100,000 Btu	nr	4.20	40.32	1219.88	189.03	1449.23
Fibre cement light quality flue pipe and fittings, joints in running lengths, pipe supports fixed to brickwork						
75mm	m	0.30	2.88	9.44	1.85	14.17
Extra over for						
bend	nr	0.35	3.36	5.98	1.40	10.74
bend with door	nr	0.35	3.36	10.46	2.07	15.89
equal tee	nr	0.45	4.32	7.49	1.77	13.5[illegible]
cone caps	nr	0.35	3.36	12.88	2.44	18.6[illegible]
GC1 terminal	nr	0.35	3.36	22.83	3.93	30.1[illegible]
GLC terminal	nr	0.35	3.36	17.82	3.18	24.3[illegible]
H piece	nr	0.25	2.40	28.73	4.67	35.8[illegible]
blank cap	nr	0.35	3.36	3.45	1.02	7.8[illegible]

	Unit	Labour hours	Net labour (£)	Net material (£)	O'heads /profit (£)	Total (£)
Fibre cement light quality flue pipe and fittings, joints in running length, pipe supports fixed to brickwork						
100mm	m	0.35	3.36	12.66	2.40	18.42
Extra over for						
bend	nr	0.40	3.84	7.46	1.70	13.00
bend with door	nr	0.40	3.84	12.20	2.41	18.45
equal tee	nr	0.45	4.32	9.29	2.04	15.65
cone caps	nr	0.40	3.84	16.84	3.10	23.78
GC1 terminal	nr	0.40	3.84	24.24	4.21	32.29
GLC terminal	nr	0.40	3.84	23.02	4.03	30.89
H piece	nr	0.45	4.32	28.73	4.96	38.01
blank cap	nr	0.40	3.84	4.35	1.23	9.42
125mm	m	0.40	3.84	16.03	2.98	22.85
Extra over for						
bend	nr	0.45	4.32	9.58	2.08	15.99
bend with door	nr	0.45	4.32	15.45	2.97	22.74
equal tee	nr	0.50	4.80	12.16	2.54	19.50
cone caps	nr	0.45	4.32	19.04	3.50	26.86
GC1 terminal	nr	0.45	4.32	27.49	4.77	36.58
GLC terminal	nr	0.45	4.32	24.75	4.36	33.43
H piece	nr	0.50	4.80	35.49	6.04	46.33
blank cap	nr	0.45	4.32	5.36	1.45	11.13
150mm	m	0.45	4.32	20.51	3.72	28.55
Extra over for						
bend	nr	0.50	4.80	11.49	2.44	18.73
bend with door	nr	0.50	4.80	17.23	3.30	25.33
equal tee	nr	0.55	5.28	14.27	2.93	22.48
cone caps	nr	0.50	4.80	22.85	4.15	31.80
GC1 terminal	nr	0.50	4.80	36.56	6.20	47.56
GLC terminal	nr	0.50	4.80	27.21	4.80	36.81
H piece	nr	0.55	5.28	43.03	7.25	55.56
blank cap	nr	0.50	4.80	7.70	1.88	14.38

	Unit	Labour hours	Net labour (£)	Net material (£)	O'heads /profit (£)	Total (£)
Fibre cement heavy quality flue pipe and fittings, joints in running length, pipe supports fixed to brickwork						
100mm	m	0.35	3.36	16.26	2.94	22.56
Extra over for						
bend	nr	0.40	3.84	10.59	2.16	16.59
bend with door	nr	0.40	3.84	17.58	3.21	24.63
equal tee	nr	0.45	4.32	10.95	2.29	17.56
H piece	nr	0.45	4.32	30.48	5.22	40.02
blank cap	nr	0.40	3.84	4.69	1.28	9.81
125mm	m	0.40	3.84	21.66	3.82	29.33
Extra over for						
bend	nr	0.45	4.32	13.11	2.61	20.04
bend with door	nr	0.45	4.32	21.66	3.90	29.88
equal tee	nr	0.50	4.80	13.70	2.77	21.28
H piece	nr	0.50	4.80	38.11	6.44	49.35
blank cap	nr	0.45	4.32	6.45	1.62	12.39
150mm	m	0.45	4.32	28.10	4.86	37.28
Extra over for						
bend	nr	0.50	4.80	15.10	2.98	22.88
bend with door	nr	0.50	4.80	25.45	4.54	34.79
equal tee	nr	0.55	5.28	16.03	3.20	24.51
H piece	nr	0.55	5.28	45.70	7.65	58.63
blank cap	nr	0.50	4.80	8.24	1.96	15.00
175mm	m	0.50	4.80	38.36	6.47	49.63
Extra over for						
bend	nr	0.55	5.28	24.04	4.40	33.72
bend with door	nr	0.55	5.28	35.23	6.08	46.59
equal tee	nr	0.60	5.76	24.60	4.55	34.91
GLC terminal	nr	0.55	5.28	37.79	6.46	49.52
H piece	nr	0.60	5.76	78.23	12.60	96.59
blank cap	nr	0.55	5.28	11.79	2.56	19.63

	Unit	Labour hours	Net labour (£)	Net material (£)	O'heads /profit (£)	Total (£)
200mm	m	0.55	5.28	41.72	7.05	54.05
Extra over for						
bend	nr	0.60	5.76	37.81	6.54	50.11
bend with door	nr	0.60	5.76	49.26	8.25	63.27
equal tee	nr	0.65	6.24	39.08	6.80	52.12
GLC terminal	nr	0.60	5.76	39.26	6.75	51.77
H piece	nr	0.65	6.24	120.73	19.05	146.02

	Unit	Material supply (£)	Material waste (%)	Total (£)
T30 RADIATORS				
MATERIALS				
Steel panel radiators, Barlo				
Single panel, 420mm high, length				
500mm	nr	15.18	2.5	15.56
650mm	nr	19.59	2.5	20.08
750mm	nr	22.50	2.5	23.06
950mm	nr	27.37	2.5	28.05
1200mm	nr	34.22	2.5	35.08
1450mm	nr	41.01	2.5	42.04
1550mm	nr	43.70	2.5	44.79
1700mm	nr	48.00	2.5	49.20
1850mm	nr	51.77	2.5	53.06
Single panel, 520mm high, length				
500mm	nr	18.41	2.5	18.87
650mm	nr	23.68	2.5	24.27
750mm	nr	27.23	2.5	27.91
850mm	nr	29.70	2.5	30.44
950mm	nr	33.05	2.5	33.88
1100mm	nr	38.10	2.5	39.05
1200mm	nr	41.44	2.5	42.48
1300mm	nr	44.67	2.5	45.79
1450mm	nr	49.62	2.5	50.86
1550mm	nr	52.95	2.5	54.27
1700mm	nr	57.80	2.5	59.25
1850mm	nr	62.75	2.5	64.32
Single panel, 620mm high, length				
500mm	nr	22.07	2.5	22.62
650mm	nr	28.53	2.5	29.24
750mm	nr	32.72	2.5	33.54
850mm	nr	35.84	2.5	36.74
950mm	nr	39.93	2.5	40.93
1100mm	nr	45.98	2.5	47.13
1200mm	nr	49.94	2.5	51.19
1300mm	nr	53.93	2.5	55.28

	Unit	Material supply (£)	Material waste (%)	Total (£)
1450mm	nr	59.95	2.5	61.45
1550mm	nr	63.93	2.5	65.53
1700mm	nr	69.85	2.5	71.60
1850mm	nr	75.67	2.5	77.56
Single panel, 720mm high, length				
500mm	nr	25.73	2.5	26.37
650mm	nr	33.15	2.5	33.98
750mm	nr	38.10	2.5	39.05
850mm	nr	41.87	2.5	42.92
950mm	nr	46.61	2.5	47.78
1100mm	nr	53.71	2.5	55.05
1200mm	nr	58.45	2.5	59.91
1300mm	nr	63.07	2.5	64.65
1450mm	nr	7.00	2.5	7.18
1550mm	nr	74.69	2.5	76.56
1700mm	nr	70.82	2.5	72.59
1850mm	nr	62.75	2.5	64.32
Single Superplus convector, 420mm high, length				
500mm	nr	22.71	2.5	23.28
650mm	nr	29.17	2.5	29.90
750mm	nr	33.38	2.5	34.21
950mm	nr	41.87	2.5	42.92
1200mm	nr	52.09	2.5	53.39
1450mm	nr	62.32	2.5	63.88
1550mm	nr	66.30	2.5	67.96
1700mm	nr	70.93	2.5	72.70
1850mm	nr	76.74	2.5	78.66
2000mm	nr	82.55	2.5	84.61
Single Superplus convector, 520mm high, length				
500mm	nr	26.37	2.5	27.03
650mm	nr	33.79	2.5	34.63
750mm	nr	38.74	2.5	39.71
850mm	nr	43.59	2.5	44.68
950mm	nr	48.43	2.5	49.64
1100mm	nr	55.65	2.5	57.04
1200mm	nr	60.38	2.5	61.89

Barlow radiators (cont'd)	Unit	Material supply (£)	Material waste (%)	Total (£)
1300mm	nr	65.12	2.5	66.75
1450mm	nr	71.46	2.5	73.25
1550mm	nr	75.23	2.5	77.11
1700mm	nr	81.80	2.5	83.85
1850mm	nr	88.47	2.5	90.68
2000mm	nr	95.25	2.5	97.63
Single Superplus convector, 620mm high, length				
500mm	nr	29.27	2.5	30.00
650mm	nr	37.57	2.5	38.51
750mm	nr	42.95	2.5	44.02
850mm	nr	48.33	2.5	49.54
950mm	nr	53.71	2.5	55.05
1100mm	nr	61.67	2.5	63.21
1200mm	nr	66.94	2.5	68.61
1300mm	nr	71.36	2.5	73.14
1450mm	nr	79.00	2.5	80.98
1550mm	nr	84.16	2.5	86.26
1700mm	nr	91.48	2.5	93.77
1850mm	nr	99.34	2.5	101.82
2000mm	nr	106.88	2.5	109.55
Single Superplus convector, 720mm high, length				
500mm	nr	33.91	2.5	34.76
650mm	nr	43.48	2.5	44.57
750mm	nr	49.84	2.5	51.09
850mm	nr	56.08	2.5	57.48
950mm	nr	62.32	2.5	63.88
1100mm	nr	69.85	2.5	71.60
1200mm	nr	75.88	2.5	77.78
1300mm	nr	81.80	2.5	83.85
1450mm	nr	90.62	2.5	92.89
1550mm	nr	96.54	2.5	98.95
1700mm	nr	105.26	2.5	107.89
1850mm	nr	115.06	2.5	117.94
2000mm	nr	122.59	2.5	125.65

	Unit	Material supply (£)	Material waste (%)	Total (£)
Double Superplus convector, 420mm high, length				
500mm	nr	43.70	2.5	44.79
650mm	nr	56.94	2.5	58.36
750mm	nr	65.65	2.5	67.29
950mm	nr	78.86	2.5	80.83
1200mm	nr	100.85	2.5	103.37
1450mm	nr	121.83	2.5	124.88
1550mm	nr	130.12	2.5	133.37
1700mm	nr	142.71	2.5	146.28
1850mm	nr	155.20	2.5	159.08
2000mm	nr	167.79	2.5	171.98
Double Superplus convector, 520mm high, length				
500mm	nr	50.80	2.5	52.07
650mm	nr	66.18	2.5	67.83
750mm	nr	76.31	2.5	78.22
850mm	nr	83.09	2.5	85.17
950mm	nr	92.89	2.5	95.21
1100mm	nr	107.52	2.5	110.21
1200mm	nr	117.31	2.5	120.24
1300mm	nr	127.00	2.5	130.18
1450mm	nr	141.63	2.5	145.17
1550mm	nr	151.32	2.5	155.10
1700mm	nr	165.86	2.5	170.01
1850mm	nr	180.38	2.5	184.89
2000mm	nr	195.02	2.5	199.90
Double Superplus convector, 620mm high, length				
500mm	nr	58.88	2.5	60.35
650mm	nr	76.63	2.5	78.55
750mm	nr	88.47	2.5	90.68
850mm	nr	96.33	2.5	98.74
950mm	nr	107.64	2.5	110.33
1100mm	nr	124.63	2.5	127.75
1200mm	nr	135.94	2.5	139.34
1300mm	nr	147.23	2.5	150.91
1450mm	nr	164.24	2.5	168.35
1550mm	nr	175.43	2.5	179.82

Barlow radiators (cont'd)	Unit	Material supply (£)	Material waste (%)	Total (£)
1700mm	nr	191.57	2.5	196.36
1850mm	nr	209.22	2.5	214.45
2000mm	nr	226.13	2.5	231.78
Double Superplus convector, 720mm high, length				
500mm	nr	66.84	2.5	68.51
650mm	nr	87.07	2.5	89.25
750mm	nr	100.52	2.5	103.03
850mm	nr	109.24	2.5	111.97
950mm	nr	121.94	2.5	124.99
1100mm	nr	141.32	2.5	144.85
1200mm	nr	154.12	2.5	157.97
1300mm	nr	166.93	2.5	171.10
1450mm	nr	186.09	2.5	190.74
1550mm	nr	198.89	2.5	203.86
1700mm	nr	218.05	2.5	223.50
Steel panel radiators, Potterton Myson				
Supaline single panel, 440mm high, length				
480mm	nr	14.90	1.0	15.05
640mm	nr	19.20	1.0	19.39
800mm	nr	23.50	1.0	23.74
960mm	nr	27.70	1.0	27.98
1120mm	nr	31.80	1.0	32.12
1280mm	nr	36.00	1.0	36.36
1440mm	nr	40.10	1.0	40.50
1600mm	nr	44.20	1.0	44.64
1760mm	nr	48.20	1.0	48.68
1920mm	nr	52.30	1.0	52.82
2080mm	nr	56.30	1.0	56.86
2240mm	nr	60.40	1.0	61.00
2400mm	nr	64.40	1.0	65.04
2560mm	nr	75.50	1.0	76.26
2720mm	nr	79.90	1.0	80.70
2880mm	nr	84.30	1.0	85.14
3040mm	nr	88.80	1.0	89.69
3200mm	nr	93.10	1.0	94.03

	Unit	Material supply (£)	Material waste (%)	Total (£)
Supaline single panel, 590mm high, length				
480mm	nr	19.50	1.0	19.70
640mm	nr	25.20	1.0	25.45
800mm	nr	30.80	1.0	31.11
960mm	nr	36.40	1.0	36.76
1120mm	nr	42.10	1.0	42.52
1280mm	nr	47.40	1.0	47.87
1440mm	nr	52.60	1.0	53.13
1600mm	nr	57.80	1.0	58.38
1760mm	nr	63.30	1.0	63.93
1920mm	nr	68.60	1.0	69.29
2080mm	nr	73.80	1.0	74.54
2240mm	nr	79.10	1.0	79.89
2400mm	nr	86.50	1.0	87.37
2560mm	nr	98.60	1.0	99.59
2720mm	nr	103.40	1.0	104.43
2880mm	nr	109.10	1.0	110.19
3040mm	nr	114.90	1.0	116.05
3200mm	nr	120.60	1.0	121.81
Supaline single panel, 740mm high, length				
480mm	nr	24.30	1.0	24.54
640mm	nr	31.50	1.0	31.82
800mm	nr	38.60	1.0	38.99
960mm	nr	45.70	1.0	46.16
1120mm	nr	54.00	1.0	54.54
1280mm	nr	61.20	1.0	61.81
1440mm	nr	68.50	1.0	69.19
1600mm	nr	75.50	1.0	76.26
1760mm	nr	82.60	1.0	83.43
1920mm	nr	89.50	1.0	90.40
2080mm	nr	93.10	1.0	94.03
2240mm	nr	100.10	1.0	101.10
2400mm	nr	106.70	1.0	107.77
2560mm	nr	120.70	1.0	121.91
2720mm	nr	127.80	1.0	129.08
2880mm	nr	134.80	1.0	136.15
3040mm	nr	141.90	1.0	143.32
3200mm	nr	149.00	1.0	150.49

Potterton Myson radiators (cont'd)	Unit	Material supply (£)	Material waste (%)	Total (£)
Supaline convector, 300mm high, length				
800mm	nr	33.80	1.0	34.14
1120mm	nr	45.80	1.0	46.26
1440mm	nr	57.60	1.0	58.18
1760mm	nr	69.20	1.0	69.89
2080mm	nr	80.90	1.0	81.71
2400mm	nr	92.10	1.0	93.02
2720mm	nr	103.70	1.0	104.74
Supaline convector, 440mm high, length				
480mm	nr	18.50	1.0	18.69
640mm	nr	24.10	1.0	24.34
800mm	nr	29.70	1.0	30.00
960mm	nr	35.20	1.0	35.55
1120mm	nr	40.80	1.0	41.21
1280mm	nr	43.30	1.0	43.73
1440mm	nr	48.40	1.0	48.88
1600mm	nr	54.50	1.0	55.05
1760mm	nr	59.70	1.0	60.30
1920mm	nr	64.80	1.0	65.45
2080mm	nr	71.50	1.0	72.22
2240mm	nr	76.60	1.0	77.37
2400mm	nr	81.90	1.0	82.72
2560mm	nr	95.10	1.0	96.05
2720mm	nr	100.78	1.0	101.79
2880mm	nr	106.40	1.0	107.46
Supaline convector, 590mm high, length				
480mm	nr	23.90	1.0	24.14
640mm	nr	31.40	1.0	31.71
800mm	nr	38.80	1.0	39.19
960mm	nr	46.10	1.0	46.56
1120mm	nr	51.90	1.0	52.42
1280mm	nr	56.80	1.0	57.37
1440mm	nr	63.80	1.0	64.44
1600mm	nr	70.50	1.0	71.21
1760mm	nr	79.50	1.0	80.30
1920mm	nr	86.50	1.0	87.37

	Unit	Material supply (£)	Material waste (%)	Total (£)
2080mm	nr	93.30	1.0	94.23
2240mm	nr	100.30	1.0	101.30
2400mm	nr	109.30	1.0	110.39
2560mm	nr	120.90	1.0	122.11
2720mm	nr	128.20	1.0	129.48
2880mm	nr	135.50	1.0	136.86
Supaline convector, 740mm high, length				
480mm	nr	29.50	1.0	29.80
640mm	nr	38.70	1.0	39.09
800mm	nr	47.90	1.0	48.38
960mm	nr	57.10	1.0	57.67
1120mm	nr	62.90	1.0	63.53
1280mm	nr	71.60	1.0	72.32
1440mm	nr	80.20	1.0	81.00
1600mm	nr	92.10	1.0	93.02
1760mm	nr	101.10	1.0	102.11
1920mm	nr	110.10	1.0	111.20
2080mm	nr	118.90	1.0	120.09
2240mm	nr	132.80	1.0	134.13
2400mm	nr	142.00	1.0	143.42
2560mm	nr	160.10	1.0	161.70
2720mm	nr	169.90	1.0	171.60
2880mm	nr	179.60	1.0	181.40
Supaline double convector, HPX one plate, 300mm high, length				
800mm	nr	49.50	1.0	50.00
1120mm	nr	68.00	1.0	68.68
1440mm	nr	86.50	1.0	87.37
1760mm	nr	104.50	1.0	105.55
2080mm	nr	122.80	1.0	124.03
2400mm	nr	140.80	1.0	142.21
2720mm	nr	158.70	1.0	160.29
Supaline double convector, HPX one plate, 440mm high, length				
480mm	nr	32.50	1.0	32.83
640mm	nr	41.20	1.0	41.61
800mm	nr	49.90	1.0	50.40

Potterton Myson radiators (cont'd)	Unit	Material supply (£)	Material waste (%)	Total (£)
960mm	nr	58.60	1.0	59.19
1120mm	nr	67.20	1.0	67.87
1280mm	nr	75.90	1.0	76.66
1440mm	nr	84.10	1.0	84.94
1600mm	nr	92.80	1.0	93.73
1760mm	nr	101.40	1.0	102.41
1920mm	nr	110.00	1.0	111.10
2080mm	nr	128.40	1.0	129.68
2240mm	nr	138.10	1.0	139.48
Supaline double convector, HPX one plate, 590mm high, length				
480mm	nr	40.20	1.0	40.60
640mm	nr	51.40	1.0	51.91
800mm	nr	62.50	1.0	63.13
960mm	nr	73.80	1.0	74.54
1120mm	nr	84.90	1.0	85.75
1280mm	nr	96.10	1.0	97.06
1440mm	nr	107.00	1.0	108.07
1600mm	nr	118.10	1.0	119.28
1760mm	nr	129.10	1.0	130.39
1920mm	nr	140.10	1.0	141.50
2080mm	nr	163.00	1.0	164.63
2240mm	nr	175.00	1.0	176.75
Supaline double convector, HPX one plate, 740mm high, length				
480mm	nr	48.50	1.0	48.99
640mm	nr	62.50	1.0	63.13
800mm	nr	76.50	1.0	77.27
960mm	nr	90.40	1.0	91.30
1120mm	nr	106.50	1.0	107.57
1280mm	nr	118.00	1.0	119.18
1440mm	nr	131.80	1.0	133.12
1600mm	nr	145.50	1.0	146.96
1760mm	nr	159.20	1.0	160.79
Supaline double convector, HDC two plates, 440mm high, length				
480mm	nr	36.80	1.0	37.17
640mm	nr	47.00	1.0	47.47

	Unit	Material supply (£)	Material waste (%)	Total (£)
800mm	nr	57.90	1.0	58.48
960mm	nr	68.40	1.0	69.08
1120mm	nr	78.80	1.0	79.59
1280mm	nr	84.50	1.0	85.35
1440mm	nr	94.50	1.0	95.45
1600mm	nr	104.40	1.0	105.44
1760mm	nr	114.40	1.0	115.54
1920mm	nr	124.30	1.0	125.54
2080mm	nr	158.20	1.0	159.78
2240mm	nr	170.10	1.0	171.80
2400mm	nr	182.10	1.0	183.92
2560mm	nr	194.10	1.0	196.04
2720mm	nr	206.10	1.0	208.16
Supaline double convector, HDC two plates, 590mm high, length				
480mm	nr	49.50	1.0	50.00
640mm	nr	64.10	1.0	64.74
800mm	nr	79.00	1.0	79.79
960mm	nr	93.60	1.0	94.54
1120mm	nr	104.90	1.0	105.95
1280mm	nr	116.30	1.0	117.46
1440mm	nr	130.20	1.0	131.50
1600mm	nr	144.10	1.0	145.54
1760mm	nr	157.80	1.0	159.38
1920mm	nr	171.70	1.0	173.42
2080mm	nr	213.20	1.0	215.33
2240mm	nr	229.40	1.0	231.69
2400mm	nr	245.60	1.0	248.06
2560mm	nr	261.70	1.0	264.32
2720mm	nr	277.90	1.0	280.68
Supaline double convector, HDC two plates, 740mm high, length				
480mm	nr	58.30	1.0	58.88
640mm	nr	75.70	1.0	76.46
800mm	nr	93.10	1.0	94.03
960mm	nr	110.70	1.0	111.81
1120mm	nr	121.40	1.0	122.61
1280mm	nr	137.90	1.0	139.28
1440mm	nr	154.40	1.0	155.94
1600mm	nr	171.00	1.0	172.71

Potterton Myson radiators (cont'd)	Unit	Material supply (£)	Material waste (%)	Total (£)
1760mm	nr	187.50	1.0	189.38
1920mm	nr	203.90	1.0	205.94
2080mm	nr	264.50	1.0	267.15
Round top single panel, 380mm high, length				
743mm	nr	16.80	1.0	16.97
946mm	nr	21.00	1.0	21.21
1149mm	nr	24.80	1.0	25.05
1352mm	nr	28.90	1.0	29.19
1556mm	nr	33.10	1.0	33.43
1759mm	nr	37.10	1.0	37.47
1962mm	nr	41.30	1.0	41.71
2165mm	nr	45.50	1.0	45.96
2638mm	nr	55.20	1.0	55.75
2572mm	nr	59.70	1.0	60.30
2775mm	nr	64.30	1.0	64.94
2978mm	nr	68.90	1.0	69.59
Round top single panel, 530mm high, length				
540mm	nr	16.70	1.0	16.87
743mm	nr	22.70	1.0	22.93
946mm	nr	28.40	1.0	28.68
1149mm	nr	33.20	1.0	33.53
1352mm	nr	39.00	1.0	39.39
1556mm	nr	44.50	1.0	44.95
1759mm	nr	50.20	1.0	50.70
1962mm	nr	55.80	1.0	56.36
2165mm	nr	61.50	1.0	62.12
2638mm	nr	74.60	1.0	75.35
2572mm	nr	80.50	1.0	81.31
2775mm	nr	86.70	1.0	87.57
2978mm	nr	92.90	1.0	93.83
Round top single panel, 685mm high, length				
540mm	nr	24.40	1.0	24.64
743mm	nr	29.20	1.0	29.49
946mm	nr	36.90	1.0	37.27
1149mm	nr	43.10	1.0	43.53

	Unit	Material supply (£)	Material waste (%)	Total (£)
1352mm	nr	50.40	1.0	50.90
1556mm	nr	57.70	1.0	58.28
1759mm	nr	65.10	1.0	65.75
1962mm	nr	72.50	1.0	73.23
2165mm	nr	71.70	1.0	72.42
2638mm	nr	96.40	1.0	97.36
2572mm	nr	104.30	1.0	105.34
2775mm	nr	112.40	1.0	113.52
2978mm	nr	120.20	1.0	121.40
Round top double panel, 380mm high, length				
743mm	nr	37.00	1.0	37.37
946mm	nr	47.00	1.0	47.47
1149mm	nr	55.20	1.0	55.75
1352mm	nr	64.80	1.0	65.45
1556mm	nr	74.40	1.0	75.14
1759mm	nr	84.20	1.0	85.04
1962mm	nr	93.50	1.0	94.44
2165mm	nr	103.20	1.0	104.23
Round top double panel, 530mm high, length				
540mm	nr	36.60	1.0	36.97
743mm	nr	50.20	1.0	50.70
946mm	nr	63.60	1.0	64.24
1149mm	nr	75.00	1.0	75.75
1352mm	nr	88.10	1.0	88.98
1556mm	nr	101.10	1.0	102.11
1759mm	nr	114.10	1.0	115.24
1962mm	nr	127.10	1.0	128.37
2165mm	nr	140.40	1.0	141.80
Round top double panel, 685mm high, length				
540mm	nr	46.60	1.0	47.07
743mm	nr	63.70	1.0	64.34
946mm	nr	81.10	1.0	81.91
1149mm	nr	95.30	1.0	96.25
1352mm	nr	112.00	1.0	113.12
1556mm	nr	128.00	1.0	129.28

Potterton Myson radiators (cont'd)	Unit	Material supply (£)	Material waste (%)	Total (£)
1759mm	nr	145.10	1.0	146.55
1962mm	nr	161.90	1.0	163.52
2165mm	nr	178.30	1.0	180.08
Round top single superconvector, 380mm high, length				
743mm	nr	21.80	1.0	22.02
946mm	nr	27.80	1.0	28.08
1047mm	nr	29.80	1.0	30.10
1149mm	nr	32.70	1.0	33.03
1251mm	nr	35.50	1.0	35.86
1352mm	nr	38.40	1.0	38.78
1454mm	nr	41.20	1.0	41.61
1556mm	nr	44.20	1.0	44.64
1759mm	nr	49.80	1.0	50.30
1962mm	nr	55.60	1.0	56.16
2165mm	nr	61.20	1.0	61.81
Round top single superconvector, 530mm high, length				
540mm	nr	21.40	1.0	21.61
641mm	nr	25.50	1.0	25.76
743mm	nr	29.50	1.0	29.80
844mm	nr	33.20	1.0	33.53
946mm	nr	37.40	1.0	37.77
1047mm	nr	40.20	1.0	40.60
1149mm	nr	43.80	1.0	44.24
1251mm	nr	47.70	1.0	48.18
1352mm	nr	51.50	1.0	52.02
1454mm	nr	55.50	1.0	56.06
1556mm	nr	59.40	1.0	59.99
1759mm	nr	67.20	1.0	67.87
1962mm	nr	74.80	1.0	75.55
2165mm	nr	82.60	1.0	83.43
Round top single superconvector, 685mm high, length				
438mm	nr	23.20	1.0	23.43
540mm	nr	27.00	1.0	27.27
641mm	nr	32.40	1.0	32.72
743mm	nr	37.00	1.0	37.37

	Unit	Material supply (£)	Material waste (%)	Total (£)
844mm	nr	42.46	1.0	42.88
946mm	nr	47.00	1.0	47.47
1047mm	nr	50.90	1.0	51.41
1149mm	nr	55.80	1.0	56.36
1352mm	nr	65.70	1.0	66.36
1556mm	nr	75.50	1.0	76.26
1759mm	nr	85.40	1.0	86.25
1962mm	nr	95.30	1.0	96.25
2165mm	nr	105.20	1.0	106.25
Round top double superconvector, 380mm high, length				
743mm	nr	44.40	1.0	44.84
946mm	nr	56.60	1.0	57.17
1047mm	nr	60.90	1.0	61.51
1149mm	nr	66.90	1.0	67.57
1251mm	nr	72.90	1.0	73.63
1352mm	nr	78.90	1.0	79.69
1454mm	nr	85.00	1.0	85.85
1556mm	nr	91.00	1.0	91.91
1759mm	nr	103.20	1.0	104.23
1962mm	nr	115.20	1.0	116.35
2165mm	nr	127.30	1.0	128.57
Round top double superconvector, 530mm high, length				
540mm	nr	43.10	1.0	43.53
641mm	nr	51.30	1.0	51.81
743mm	nr	59.50	1.0	60.10
844mm	nr	67.70	1.0	68.38
946mm	nr	76.20	1.0	76.96
1047mm	nr	82.10	1.0	82.92
1149mm	nr	90.30	1.0	91.20
1251mm	nr	98.30	1.0	99.28
1352mm	nr	106.40	1.0	107.46
1454mm	nr	114.60	1.0	115.75
1556mm	nr	122.70	1.0	123.93
1759mm	nr	138.80	1.0	140.19
1962mm	nr	155.20	1.0	156.75
2165mm	nr	171.90	1.0	173.62

Potterton Myson radiators (cont'd)	Unit	Material supply (£)	Material waste (%)	Total (£)
Round top double superconvector, 685mm high, length				
438mm	nr	41.00	1.0	41.41
540mm	nr	53.70	1.0	54.24
641mm	nr	63.90	1.0	64.54
743mm	nr	74.30	1.0	75.04
844mm	nr	84.60	1.0	85.45
946mm	nr	95.00	1.0	95.95
1047mm	nr	102.20	1.0	103.22
1149mm	nr	112.50	1.0	113.63
1352mm	nr	132.60	1.0	133.93
1556mm	nr	153.00	1.0	154.53
1759mm	nr	173.20	1.0	174.93
1962mm	nr	193.80	1.0	195.74
2165mm	nr	214.20	1.0	216.34

	Unit	Labour hours	Net labour (£)	Net material (£)	O'heads /profit (£)	Total (£)
MEDIUM TEMPERATURE HOT WATER HEATING RADIATORS						
Steel panel radiators, Barlo, plugged and screwed to brickwork with concealed brackets, satin primer finish						
Single panel, 420mm high, length						
500mm	nr	1.00	9.60	15.56	0.63	25.79
650mm	nr	1.00	9.60	20.08	0.74	30.42
750mm	nr	1.05	10.08	23.06	0.83	33.97
950mm	nr	1.10	10.56	28.05	0.97	39.58
1200mm	nr	1.15	11.04	35.08	1.15	47.27
1450mm	nr	1.20	11.52	42.04	1.34	54.90
1550mm	nr	1.25	12.00	44.79	1.42	58.21
1700mm	nr	1.30	12.48	49.20	1.54	63.22
1850mm	nr	1.35	12.96	53.06	1.65	67.67
Single panel, 520mm high, length						
500mm	nr	1.00	9.60	18.59	0.70	28.89
650mm	nr	1.00	9.60	23.92	0.84	34.36
750mm	nr	1.05	10.08	27.50	0.94	38.52
850mm	nr	1.10	10.56	30.00	1.01	41.57
950mm	nr	1.10	10.56	33.38	1.10	45.04
1100mm	nr	1.15	11.04	38.48	1.24	50.76
1200mm	nr	1.15	11.04	41.85	1.32	54.21
1300mm	nr	1.20	11.52	45.12	1.42	58.06
1450mm	nr	1.20	11.52	50.12	1.54	63.18
1550mm	nr	1.25	12.00	53.48	1.64	67.12
1700mm	nr	1.30	12.48	58.38	1.77	72.63
1850mm	nr	1.35	12.96	63.38	1.91	78.25
Single panel, 620mm high, length						
500mm	nr	1.10	10.56	22.29	0.82	33.67
650mm	nr	1.10	10.56	28.82	0.98	40.36
750mm	nr	1.15	11.04	33.05	1.10	45.19
850mm	nr	1.20	11.52	36.20	1.19	48.91
950mm	nr	1.20	11.52	40.33	1.30	53.15
1100mm	nr	1.25	12.00	46.44	1.46	59.90
1200mm	nr	1.25	12.00	50.44	1.56	64.00

Barlo radiators (cont'd)	Unit	Labour hours	Net labour (£)	Net material (£)	O'heads /profit (£)	Total (£)
1300mm	nr	1.30	12.48	54.47	1.67	68.62
1450mm	nr	1.30	12.48	60.55	1.83	74.86
1550mm	nr	1.35	12.96	64.57	1.94	79.47
1700mm	nr	1.40	13.44	70.55	2.10	86.09
1850mm	nr	1.45	13.92	76.43	2.26	92.61
Single panel, 720mm high, length						
500mm	nr	1.10	10.56	25.99	0.91	37.46
650mm	nr	1.10	10.56	33.48	1.10	45.14
750mm	nr	1.15	11.04	38.48	1.24	50.76
850mm	nr	1.20	11.52	42.29	1.35	55.16
950mm	nr	1.20	11.52	47.08	1.47	60.07
1100mm	nr	1.25	12.00	54.25	1.66	67.91
1200mm	nr	1.25	12.00	59.03	1.78	72.81
1300mm	nr	1.30	12.48	63.70	1.90	78.08
1450mm	nr	1.30	12.48	69.04	2.04	83.56
1550mm	nr	1.35	12.96	75.44	2.21	90.61
1700mm	nr	1.40	13.44	71.53	2.12	87.09
1850mm	nr	1.45	13.92	63.38	1.93	79.23
Single Superplus convector, 420mm high, length						
500mm	nr	1.00	9.60	22.94	0.81	33.35
650mm	nr	1.00	9.60	29.46	0.98	40.04
750mm	nr	1.05	10.08	33.71	1.09	44.88
950mm	nr	1.10	10.56	42.29	1.32	54.17
1200mm	nr	1.15	11.04	52.61	1.59	65.24
1450mm	nr	1.20	11.52	62.94	1.86	76.32
1550mm	nr	1.25	12.00	66.96	1.97	80.93
1700mm	nr	1.30	12.48	71.64	2.10	86.22
1850mm	nr	1.35	12.96	77.51	2.26	92.73
2000mm	nr	1.40	13.44	83.38	2.42	99.24
Single Superplus convector, 520mm high, length						
500mm	nr	1.00	9.60	26.63	0.91	37.14
650mm	nr	1.00	9.60	34.13	1.09	44.82
750mm	nr	1.05	10.08	39.13	1.23	50.44
850mm	nr	1.10	10.56	44.03	1.36	55.95
950mm	nr	1.10	10.56	48.91	1.49	60.96
1100mm	nr	1.15	11.04	56.21	1.68	68.93

	Unit	Labour hours	Net labour (£)	Net material (£)	O'heads /profit (£)	Total (£)
1200mm	nr	1.15	11.04	60.98	1.80	73.82
1300mm	nr	1.20	11.52	65.77	1.93	79.22
1450mm	nr	1.20	11.52	72.17	2.09	85.78
1550mm	nr	1.25	12.00	75.98	2.20	90.18
1700mm	nr	1.30	12.48	82.62	2.38	97.48
1850mm	nr	1.35	12.96	89.35	2.56	104.87
2000mm	nr	1.40	13.44	96.20	2.74	112.38
Single Superplus convector, 620mm high, length						
500mm	nr	1.10	10.56	29.56	1.00	41.12
650mm	nr	1.10	10.56	37.95	1.21	49.72
750mm	nr	1.15	11.04	43.38	1.36	55.78
850mm	nr	1.20	11.52	48.81	1.51	61.84
950mm	nr	1.20	11.52	54.25	1.64	67.41
1100mm	nr	1.25	12.00	62.29	1.86	76.15
1200mm	nr	1.25	12.00	67.61	1.99	81.60
1300mm	nr	1.30	12.48	72.07	2.11	86.66
1450mm	nr	1.30	12.48	79.79	2.31	94.58
1550mm	nr	1.35	12.96	85.00	2.45	100.41
1700mm	nr	1.40	13.44	92.39	2.65	108.48
1850mm	nr	1.45	13.92	100.33	2.86	117.11
2000mm	nr	1.50	14.40	107.95	3.06	125.41
Single Superplus convector, 720mm high, length						
500mm	nr	1.10	10.56	34.25	1.12	45.93
650mm	nr	1.10	10.56	43.91	1.36	55.83
750mm	nr	1.15	11.04	50.34	1.53	62.91
850mm	nr	1.20	11.52	56.64	1.70	69.86
950mm	nr	1.20	11.52	62.94	1.86	76.32
1100mm	nr	1.25	12.00	70.55	2.06	84.61
1200mm	nr	1.25	12.00	76.64	2.22	90.86
1300mm	nr	1.30	12.48	82.62	2.38	97.48
1450mm	nr	1.30	12.48	91.53	2.60	106.61
1550mm	nr	1.35	12.96	97.51	2.76	113.23
1700mm	nr	1.40	13.44	106.31	2.99	122.74
1850mm	nr	1.45	13.92	116.21	3.25	133.38
2000mm	nr	1.50	14.40	123.82	3.46	141.68

Barlo radiators (cont'd)	Unit	Labour hours	Net labour (£)	Net material (£)	O'heads /profit (£)	Total (£)
Double Superplus convector, 420mm high, length						
500mm	nr	1.00	9.60	44.14	1.34	55.08
650mm	nr	1.00	9.60	57.51	1.68	68.79
750mm	nr	1.05	10.08	66.31	1.91	78.30
950mm	nr	1.10	10.56	79.65	2.26	92.47
1200mm	nr	1.15	11.04	101.86	2.82	115.72
1450mm	nr	1.20	11.52	123.05	3.36	137.93
1550mm	nr	1.25	12.00	131.42	3.59	147.01
1700mm	nr	1.30	12.48	144.14	3.92	160.54
1850mm	nr	1.35	12.96	156.75	4.24	173.95
2000mm	nr	1.40	13.44	169.47	4.57	187.48
Double Superplus convector, 520mm high, length						
500mm	nr	1.00	9.60	51.31	1.52	62.43
650mm	nr	1.00	9.60	66.84	1.91	78.35
750mm	nr	1.05	10.08	77.07	2.18	89.33
850mm	nr	1.10	10.56	83.92	2.36	96.84
950mm	nr	1.10	10.56	93.82	2.61	106.99
1100mm	nr	1.15	11.04	108.60	2.99	122.63
1200mm	nr	1.15	11.04	118.48	3.24	132.76
1300mm	nr	1.20	11.52	128.27	3.49	143.28
1450mm	nr	1.20	11.52	143.05	3.86	158.43
1550mm	nr	1.25	12.00	152.83	4.12	168.95
1700mm	nr	1.30	12.48	167.52	4.50	184.50
1850mm	nr	1.35	12.96	182.18	4.88	200.02
2000mm	nr	1.40	13.44	196.97	5.26	215.67
Double Superplus convector, 620mm high, length						
500mm	nr	1.10	10.56	59.47	1.75	71.78
650mm	nr	1.10	10.56	77.40	2.20	90.16
750mm	nr	1.15	11.04	89.35	2.51	102.90
850mm	nr	1.20	11.52	97.29	2.72	111.53
950mm	nr	1.20	11.52	108.72	3.01	123.25
1100mm	nr	1.25	12.00	125.88	3.45	141.33
1200mm	nr	1.25	12.00	137.30	3.73	153.03
1300mm	nr	1.30	12.48	148.70	4.03	165.21
1450mm	nr	1.30	12.48	165.88	4.46	182.82
1550mm	nr	1.35	12.96	177.18	4.75	194.89

	Unit	Labour hours	Net labour (£)	Net material (£)	O'heads /profit (£)	Total (£)
1700mm	nr	1.40	13.44	193.49	5.17	212.10
1850mm	nr	1.45	13.92	211.31	5.63	230.86
2000mm	nr	1.50	14.40	228.39	6.07	248.86
Double Superplus convector, 720mm high, length						
500mm	nr	1.00	9.60	67.51	1.93	79.04
650mm	nr	1.00	9.60	87.94	2.44	99.98
750mm	nr	1.05	10.08	101.53	2.79	114.40
850mm	nr	1.10	10.56	110.33	3.02	123.91
950mm	nr	1.10	10.56	123.16	3.34	137.06
1100mm	nr	1.15	11.04	142.73	3.84	157.61
1200mm	nr	1.15	11.04	155.66	4.17	170.87
1300mm	nr	1.20	11.52	168.60	4.50	184.62
1450mm	nr	1.20	11.52	187.95	4.99	204.46
1550mm	nr	1.25	12.00	200.88	5.32	218.20
1700mm	nr	1.30	12.48	220.23	5.82	238.53
Steel panel radiators, Potterton Myson, plugged and screwed to brickwork with concealed brackets, full finish in white						
Supaline single panel, 440mm high, length						
480mm	nr	1.00	9.60	15.05	0.62	25.27
640mm	nr	1.00	9.60	19.39	0.72	29.71
800mm	nr	1.00	9.60	23.74	0.83	34.17
960mm	nr	1.10	10.56	27.98	0.96	39.50
1120mm	nr	1.10	10.56	32.12	1.07	43.75
1280mm	nr	1.15	11.04	36.36	1.19	48.59
1440mm	nr	1.20	11.52	40.50	1.30	53.32
1600mm	nr	1.25	12.00	44.64	1.42	58.06
1760mm	nr	1.30	12.48	48.68	1.53	62.69
1920mm	nr	1.35	12.96	52.82	1.64	67.42
2080mm	nr	1.40	13.44	56.86	1.76	72.06
2240mm	nr	1.45	13.92	61.00	1.87	76.79
2400mm	nr	1.50	14.40	65.04	1.99	81.43
2560mm	nr	1.60	15.36	76.26	2.29	93.91
2720mm	nr	1.80	17.28	80.70	2.45	100.43

Potterton Myson radiators (cont'd)	Unit	Labour hours	Net labour (£)	Net material (£)	O'heads /profit (£)	Total (£)
2880mm	nr	1.90	18.24	85.14	2.58	105.96
3040mm	nr	2.20	21.12	89.69	2.77	113.58
3200mm	nr	2.50	24.00	94.03	2.95	120.98
Supaline single panel, 590mm high, length						
480mm	nr	1.00	9.60	19.70	0.73	30.03
640mm	nr	1.00	9.60	25.45	0.88	35.93
800mm	nr	1.05	10.08	31.11	1.03	42.22
960mm	nr	1.10	10.56	36.76	1.18	48.50
1120mm	nr	1.10	10.56	42.52	1.33	54.41
1280mm	nr	1.15	11.04	47.87	1.47	60.38
1440mm	nr	1.20	11.52	53.13	1.62	66.27
1600mm	nr	1.25	12.00	58.38	1.76	72.14
1760mm	nr	1.30	12.48	63.93	1.91	78.32
1920mm	nr	1.35	12.96	69.29	2.06	84.31
2080mm	nr	1.40	13.44	74.54	2.20	90.18
2240mm	nr	1.45	13.92	79.89	2.35	96.16
2400mm	nr	1.50	14.40	87.37	2.54	104.31
2560mm	nr	1.60	15.36	99.59	2.87	117.82
2720mm	nr	1.90	18.24	104.43	3.07	125.74
2880mm	nr	1.90	18.24	110.19	3.21	131.64
3040mm	nr	2.20	21.12	116.05	3.43	140.60
3200mm	nr	2.50	24.00	121.81	3.65	149.46
Supaline single panel, 740mm high, length						
480mm	nr	1.10	10.56	24.54	0.88	35.98
640mm	nr	1.10	10.56	31.82	1.06	43.44
800mm	nr	1.20	11.52	38.99	1.26	51.77
960mm	nr	1.20	11.52	46.16	1.44	59.12
1120mm	nr	1.25	12.00	54.54	1.66	68.20
1280mm	nr	1.25	12.00	61.81	1.85	75.66
1440mm	nr	1.30	12.48	69.19	2.04	83.71
1600mm	nr	1.35	12.96	76.26	2.23	91.45
1760mm	nr	1.40	13.44	83.43	2.42	99.29
1920mm	nr	1.45	13.92	90.40	2.61	106.93
2080mm	nr	1.50	14.40	94.03	2.71	111.14
2240mm	nr	1.55	14.88	101.10	2.90	118.88
2400mm	nr	1.60	15.36	107.77	3.08	126.21
2560mm	nr	1.70	16.32	121.91	3.46	141.69
2720mm	nr	1.90	18.24	129.08	3.68	151.00

	Unit	Labour hours	Net labour (£)	Net material (£)	O'heads /profit (£)	Total (£)
2880mm	nr	2.00	19.20	136.15	3.88	159.23
3040mm	nr	2.30	22.08	143.32	4.14	169.54
3200mm	nr	2.60	24.96	150.49	4.39	179.84
Supaline convector, 300mm high, length						
800mm	nr	1.00	9.60	34.14	1.09	44.83
1120mm	nr	1.10	10.56	46.26	1.42	58.24
1440mm	nr	1.20	11.52	58.18	1.74	71.44
1760mm	nr	1.30	12.48	69.89	2.06	84.43
2080mm	nr	1.40	13.44	81.71	2.38	97.53
2400mm	nr	1.50	14.40	93.02	2.69	110.11
2720mm	nr	1.80	17.28	104.74	3.05	125.07
Supaline convector, 440mm high, length						
480mm	nr	1.00	9.60	18.69	0.71	29.00
640mm	nr	1.00	9.60	24.34	0.85	34.79
800mm	nr	1.05	10.08	30.00	1.00	41.08
960mm	nr	1.10	10.56	35.55	1.15	47.26
1120mm	nr	1.10	10.56	41.21	1.29	53.06
1280mm	nr	1.15	11.04	43.73	1.37	56.14
1440mm	nr	1.20	11.52	48.88	1.51	61.91
1600mm	nr	1.25	0.00	54.94	1.37	56.31
1760mm	nr	1.30	0.00	60.30	1.51	61.81
1920mm	nr	1.35	12.96	65.45	1.96	80.37
2080mm	nr	1.40	13.44	72.22	2.14	87.80
2240mm	nr	1.45	13.92	77.37	2.28	93.57
2400mm	nr	1.50	14.40	82.72	2.43	99.55
2560mm	nr	1.60	15.36	96.05	2.79	114.20
2720mm	nr	1.80	17.28	101.79	2.98	122.05
2880mm	nr	1.90	18.24	107.46	3.14	128.84
Supaline convector, 590mm high, length						
480mm	nr	1.00	9.60	24.14	0.84	34.58
640mm	nr	1.00	9.60	31.71	1.03	42.34
800mm	nr	1.05	10.08	39.19	1.23	50.50
960mm	nr	1.10	10.56	46.56	1.43	58.55
1120mm	nr	1.10	10.56	52.42	1.57	64.55
1280mm	nr	1.15	11.04	57.37	1.71	70.12

Potterton Myson radiators (cont'd)	Unit	Labour hours	Net labour (£)	Net material (£)	O'heads /profit (£)	Total (£)
1440mm	nr	1.20	11.52	64.44	1.90	77.86
1600mm	nr	1.25	12.00	71.21	2.08	85.29
1760mm	nr	1.30	12.48	80.30	2.32	95.10
1920mm	nr	1.35	12.96	87.37	2.51	102.84
2080mm	nr	1.40	13.44	94.23	2.69	110.36
2240mm	nr	1.45	13.92	101.30	2.88	118.10
2400mm	nr	1.50	14.40	110.39	3.12	127.91
2560mm	nr	1.60	15.36	122.11	3.44	140.91
2720mm	nr	1.80	17.28	129.48	3.67	150.43
2880mm	nr	1.90	18.24	136.86	3.88	158.98
Supaline convector, 740mm high, length						
480mm	nr	1.10	10.56	29.80	1.01	41.37
640mm	nr	1.10	10.56	39.09	1.24	50.89
800mm	nr	1.15	11.04	48.38	1.49	60.91
960mm	nr	1.20	11.52	57.67	1.73	70.92
1120mm	nr	1.20	11.52	63.53	1.88	76.93
1280mm	nr	1.25	12.00	72.32	2.11	86.43
1440mm	nr	1.30	12.48	81.00	2.34	95.82
1600mm	nr	1.35	12.96	93.02	2.65	108.63
1760mm	nr	1.40	13.44	102.11	2.89	118.44
1920mm	nr	1.45	13.92	111.20	3.13	128.25
2080mm	nr	1.50	14.40	120.09	3.36	137.85
2240mm	nr	1.55	14.88	134.13	3.73	152.74
2400mm	nr	1.60	15.36	143.42	3.97	162.75
2560mm	nr	1.70	16.32	161.70	4.45	182.47
2720mm	nr	1.80	17.28	171.60	4.72	193.60
2880mm	nr	1.90	18.24	181.40	4.99	204.63
Supaline double convector, HPX one plate, 300mm high, length						
800mm	nr	1.25	12.00	50.00	1.55	63.55
1120mm	nr	1.30	12.48	68.68	2.03	83.19
1440mm	nr	1.40	13.44	87.37	2.52	103.33
1760mm	nr	1.50	14.40	105.55	3.00	122.95
2080mm	nr	1.60	15.36	124.03	3.48	142.87
2400mm	nr	1.70	16.32	142.21	3.96	162.49
2720mm	nr	2.10	20.16	160.29	4.51	184.96

	Unit	Labour hours	Net labour (£)	Net material (£)	O'heads /profit (£)	Total (£)
Supaline double convector, HPX one plate, 440mm high, length						
480mm	nr	1.30	12.48	32.83	1.13	46.44
640mm	nr	1.30	12.48	41.61	1.35	55.44
800mm	nr	1.35	12.96	50.40	1.58	64.94
960mm	nr	1.40	13.44	59.19	1.82	74.45
1120mm	nr	1.40	13.44	67.87	2.03	83.34
1280mm	nr	1.45	13.92	76.66	2.26	92.84
1440mm	nr	1.50	14.40	84.94	2.48	101.82
1600mm	nr	1.55	14.88	93.73	2.72	111.33
1760mm	nr	1.60	15.36	102.41	2.94	120.71
1920mm	nr	1.65	15.84	111.10	3.17	130.11
2080mm	nr	1.70	16.32	129.68	3.65	149.65
2240mm	nr	1.75	16.80	139.48	3.91	160.19
Supaline double convector, HPX one plate, 590mm high, length						
480mm	nr	1.30	12.48	40.60	1.33	54.41
640mm	nr	1.30	12.48	51.91	1.61	66.00
800mm	nr	1.35	12.96	63.13	1.90	77.99
960mm	nr	1.40	13.44	74.54	2.20	90.18
1120mm	nr	1.40	13.44	85.75	2.48	101.67
1280mm	nr	1.45	13.92	97.06	2.77	113.75
1440mm	nr	1.50	14.40	108.07	3.06	125.53
1600mm	nr	1.55	14.88	119.28	3.35	137.51
1760mm	nr	1.60	15.36	130.39	3.64	149.39
1920mm	nr	1.65	15.84	141.50	3.93	161.27
2080mm	nr	1.70	16.32	164.63	4.52	185.47
2240mm	nr	1.75	16.80	176.75	4.84	198.39
Supaline double convector, HPX one plate, 740mm high, length						
480mm	nr	1.40	13.44	48.99	1.56	63.99
640mm	nr	1.40	13.44	63.13	1.91	78.48
800mm	nr	1.45	13.92	77.27	2.28	93.47
960mm	nr	1.50	14.40	91.30	2.64	108.34
1120mm	nr	1.50	14.40	107.57	3.05	125.02
1280mm	nr	1.55	14.88	119.18	3.35	137.41
1440mm	nr	1.60	15.36	133.12	3.71	152.19
1600mm	nr	1.65	15.84	146.96	4.07	166.87
1760mm	nr	1.70	16.32	160.79	4.43	181.54

Potterton Myson radiators (cont'd)	Unit	Labour hours	Net labour (£)	Net material (£)	O'heads /profit (£)	Total (£)
Supaline double convector, HDC two plates, 440mm high, length						
480mm	nr	1.30	12.48	37.17	1.24	50.89
640mm	nr	1.30	12.48	47.47	1.50	61.45
800mm	nr	1.35	12.96	58.48	1.79	73.23
960mm	nr	1.40	13.44	69.08	2.06	84.58
1120mm	nr	1.40	13.44	79.59	2.33	95.36
1280mm	nr	1.45	13.92	85.35	2.48	101.75
1440mm	nr	1.50	14.40	95.45	2.75	112.60
1600mm	nr	1.55	14.88	105.44	3.01	123.33
1760mm	nr	1.60	15.36	115.54	3.27	134.17
1920mm	nr	1.65	15.84	125.54	3.53	144.91
2080mm	nr	1.70	16.32	159.78	4.40	180.50
2240mm	nr	1.75	16.80	171.80	4.71	193.32
2400mm	nr	1.80	17.28	183.92	5.03	206.23
2560mm	nr	1.90	18.24	196.04	5.36	219.64
2720mm	nr	2.10	20.16	208.16	5.71	234.03
Supaline double convector, HDC two plates, 590mm high, length						
480mm	nr	1.30	12.48	50.00	1.56	64.04
640mm	nr	1.30	12.48	64.74	1.93	79.15
800mm	nr	1.35	12.96	79.79	2.32	95.07
960mm	nr	1.40	13.44	94.54	2.70	110.68
1120mm	nr	1.40	13.44	105.95	2.98	122.37
1280mm	nr	1.45	13.92	117.46	3.28	134.66
1440mm	nr	1.50	14.40	131.50	3.65	149.55
1600mm	nr	1.55	14.88	145.54	4.01	164.43
1760mm	nr	1.60	15.36	159.38	4.37	179.11
1920mm	nr	1.65	15.84	173.42	4.73	193.99
2080mm	nr	1.70	16.32	215.33	5.79	237.44
2240mm	nr	1.75	16.80	231.69	6.21	254.70
2400mm	nr	1.80	17.28	248.06	6.63	271.97
2560mm	nr	1.90	18.24	264.32	7.06	289.62
2720mm	nr	2.10	20.16	280.68	7.52	308.36
Supaline double convector, HDC two plates, 740mm high, length						
480mm	nr	1.40	13.44	58.88	1.81	74.13
640mm	nr	1.40	13.44	76.46	2.25	92.15
800mm	nr	1.45	13.92	94.03	2.70	110.65

	Unit	Labour hours	Net labour (£)	Net material (£)	O'heads /profit (£)	Total (£)
960mm	nr	1.50	14.40	111.81	3.16	129.37
1120mm	nr	1.50	14.40	122.61	3.43	140.44
1280mm	nr	1.55	14.88	139.28	3.85	158.01
1440mm	nr	1.60	15.36	155.94	4.28	175.58
1600mm	nr	1.65	15.84	172.71	4.71	193.26
1760mm	nr	1.70	16.32	189.38	5.14	210.84
1920mm	nr	1.75	16.80	205.94	5.57	228.31
2080mm	nr	1.80	17.28	267.15	7.11	291.54
Round top single panel, 380mm high, length						
743mm	nr	1.05	10.08	16.97	0.68	27.73
946mm	nr	1.10	10.56	21.21	0.79	32.56
1149mm	nr	1.15	11.04	25.05	0.90	36.99
1352mm	nr	1.20	11.52	29.19	1.02	41.73
1556mm	nr	1.25	12.00	33.43	1.14	46.57
1759mm	nr	1.30	12.48	37.47	1.25	51.20
1962mm	nr	1.35	12.96	41.71	1.37	56.04
2165mm	nr	1.40	13.44	45.96	1.49	60.89
2638mm	nr	1.60	15.36	55.75	1.78	72.89
2572mm	nr	1.70	16.32	60.30	1.92	78.54
2775mm	nr	1.80	17.28	64.94	2.06	84.28
2978mm	nr	2.00	19.20	69.59	2.22	91.01
Round top single panel, 580mm high, length						
540mm	nr	1.00	9.60	16.87	0.66	27.13
743mm	nr	1.05	10.08	22.93	0.83	33.84
946mm	nr	1.10	10.56	28.68	0.98	40.22
1149mm	nr	1.15	11.04	33.53	1.11	45.68
1352mm	nr	1.20	11.52	39.39	1.27	52.18
1556mm	nr	1.25	12.00	44.95	1.42	58.37
1759mm	nr	1.30	12.48	50.70	1.58	64.76
1962mm	nr	1.35	12.96	56.36	1.73	71.05
2165mm	nr	1.40	13.44	62.12	1.89	77.45
2638mm	nr	1.60	15.36	75.35	2.27	92.98
2572mm	nr	1.70	16.32	81.31	2.44	100.07
2775mm	nr	1.80	17.28	87.57	2.62	107.47
2978mm	nr	2.00	19.20	93.83	2.83	115.86

Potterton Myson radiators (cont'd)	Unit	Labour hours	Net labour (£)	Net material (£)	O'heads /profit (£)	Total (£)
Round top single panel, 685mm high, length						
540mm	nr	1.05	10.08	24.64	0.87	35.59
743mm	nr	1.10	10.56	29.49	1.00	41.05
946mm	nr	1.15	11.04	37.27	1.21	49.52
1149mm	nr	1.20	11.52	43.53	1.38	56.43
1352mm	nr	1.25	12.00	50.90	1.57	64.47
1556mm	nr	1.30	12.48	58.28	1.77	72.53
1759mm	nr	1.35	12.96	65.75	1.97	80.68
1962mm	nr	1.40	13.44	73.23	2.17	88.84
2165mm	nr	1.45	13.92	72.42	2.16	88.50
2638mm	nr	1.60	15.36	97.36	2.82	115.54
2572mm	nr	1.75	16.80	105.34	3.05	125.19
2775mm	nr	1.85	17.76	113.52	3.28	134.56
2978mm	nr	2.05	19.68	121.40	3.53	144.61
Round top double panel, 380mm high, length						
743mm	nr	1.15	11.04	37.37	1.21	49.62
946mm	nr	1.20	11.52	47.47	1.47	60.46
1149mm	nr	1.30	12.48	55.75	1.71	69.94
1352mm	nr	1.40	13.44	65.45	1.97	80.86
1556mm	nr	1.50	14.40	75.14	2.24	91.78
1759mm	nr	1.70	16.32	85.04	2.53	103.89
1962mm	nr	1.80	17.28	94.44	2.79	114.51
2165mm	nr	1.90	18.24	104.23	3.06	125.53
Round top double panel, 530mm high, length						
540mm	nr	1.10	10.56	36.97	1.19	48.72
743mm	nr	1.15	11.04	50.70	1.54	63.28
946mm	nr	1.20	11.52	64.24	1.89	77.65
1149mm	nr	1.25	12.00	75.75	2.19	89.94
1352mm	nr	1.30	12.48	88.98	2.54	104.00
1556mm	nr	1.40	13.44	102.11	2.89	118.44
1759mm	nr	1.45	13.92	115.24	3.23	132.39
1962mm	nr	1.50	14.40	128.37	3.57	146.34
2165mm	nr	1.70	16.32	141.80	3.95	162.07

RADIATORS

	Unit	Labour hours	Net labour (£)	Net material (£)	O'heads /profit (£)	Total (£)
Round top double panel, 685mm high, length						
540mm	nr	1.20	11.52	47.07	1.46	60.05
743mm	nr	1.25	12.00	64.34	1.91	78.25
946mm	nr	1.30	12.48	81.91	2.36	96.75
1149mm	nr	1.40	13.44	96.25	2.74	112.43
1352mm	nr	1.45	13.92	113.12	3.18	130.22
1556mm	nr	1.50	14.40	129.28	3.59	147.27
1759mm	nr	1.55	14.88	146.55	4.04	165.47
1962mm	nr	1.60	15.36	163.52	4.47	183.35
2165mm	nr	1.80	17.28	180.08	4.93	202.29
Round top single superconvector, 380mm high, length						
743mm	nr	1.05	10.08	22.02	0.80	32.90
946mm	nr	1.10	10.56	28.08	0.97	39.61
1047mm	nr	1.15	11.04	30.10	1.03	42.17
1149mm	nr	1.15	11.04	33.03	1.10	45.17
1251mm	nr	1.20	11.52	35.86	1.18	48.56
1352mm	nr	1.20	11.52	38.78	1.26	51.56
1454mm	nr	1.25	12.00	41.61	1.34	54.95
1556mm	nr	1.25	12.00	44.64	1.42	58.06
1759mm	nr	1.30	12.48	50.30	1.57	64.35
1962mm	nr	1.35	12.96	56.16	1.73	70.85
2165mm	nr	1.40	13.44	61.81	1.88	77.13
Round top single superconvector, 530mm high, length						
540mm	nr	1.00	9.60	21.61	0.78	31.99
641mm	nr	1.00	9.60	25.76	0.88	36.24
743mm	nr	1.05	10.08	29.80	1.00	40.88
844mm	nr	1.05	10.08	33.53	1.09	44.70
946mm	nr	1.10	10.56	37.77	1.21	49.54
1047mm	nr	1.15	11.04	40.60	1.29	52.93
1149mm	nr	1.15	11.04	44.24	1.38	56.66
1251mm	nr	1.20	11.52	48.18	1.49	61.19
1352mm	nr	1.20	11.52	52.02	1.59	65.13
1454mm	nr	1.25	12.00	56.06	1.70	69.76

Potterton Myson radiators (cont'd)	Unit	Labour hours	Net labour (£)	Net material (£)	O'heads /profit (£)	Total (£)
1556mm	nr	1.25	12.00	59.99	1.80	73.79
1759mm	nr	1.30	12.48	67.87	2.01	82.36
1962mm	nr	1.35	12.96	75.55	2.21	90.72
2165mm	nr	1.40	13.44	83.43	2.42	99.29
Round top single superconvector, 685mm high, length						
438mm	nr	1.10	10.56	23.43	0.85	34.84
540mm	nr	1.10	10.56	27.27	0.95	38.78
641mm	nr	1.15	11.04	32.72	1.09	44.85
743mm	nr	1.15	11.04	37.37	1.21	49.62
844mm	nr	1.15	11.04	42.88	1.35	55.27
946mm	nr	1.20	11.52	47.47	1.47	60.46
1047mm	nr	1.25	12.00	51.41	1.59	65.00
1149mm	nr	1.25	12.00	56.36	1.71	70.07
1352mm	nr	1.30	12.48	66.36	1.97	80.81
1556mm	nr	1.35	12.96	76.26	2.23	91.45
1759mm	nr	1.40	13.44	86.25	2.49	102.18
1962mm	nr	1.45	13.92	96.25	2.75	112.92
2165mm	nr	1.50	14.40	106.25	3.02	123.67
Round top double superconvector, 380mm high, length						
743mm	nr	1.15	11.04	44.84	1.40	57.28
946mm	nr	1.20	11.52	57.17	1.72	70.41
1047mm	nr	1.25	12.00	61.51	1.84	75.35
1149mm	nr	1.25	12.00	61.51	1.84	75.35
1251mm	nr	1.30	12.48	73.63	2.15	88.26
1352mm	nr	1.30	12.48	79.69	2.30	94.47
1454mm	nr	1.35	12.96	85.85	2.47	101.28
1556mm	nr	1.35	12.96	91.91	2.62	107.49
1759mm	nr	1.40	13.44	104.23	2.94	120.61
1962mm	nr	1.45	13.92	116.35	3.26	133.53
2165mm	nr	1.50	14.40	128.57	3.57	146.54
Round top double superconvector, 530mm high, length						
540mm	nr	1.00	9.60	43.53	1.33	54.4[illegible]
641mm	nr	1.10	10.56	51.81	1.56	63.9[illegible]
743mm	nr	1.15	11.04	60.10	1.78	72.9[illegible]
844mm	nr	1.15	11.04	68.38	1.99	81.4[illegible]

	Unit	Labour hours	Net labour (£)	Net material (£)	O'heads /profit (£)	Total (£)
946mm	nr	1.20	11.52	76.96	2.21	90.69
1047mm	nr	1.25	12.00	82.92	2.37	97.29
1149mm	nr	1.25	12.00	91.20	2.58	105.78
1251mm	nr	1.30	12.48	99.28	2.79	114.55
1352mm	nr	1.30	12.48	107.46	3.00	122.94
1454mm	nr	1.35	12.96	115.75	3.22	131.93
1556mm	nr	1.35	12.96	123.93	3.42	140.31
1759mm	nr	1.40	13.44	140.19	3.84	157.47
1962mm	nr	1.45	13.92	156.75	4.27	174.94
2165mm	nr	1.50	14.40	173.62	4.70	192.72
Round top double superconvector, 685mm high, length						
438mm	nr	1.10	10.56	41.41	1.30	53.27
540mm	nr	1.10	10.56	54.24	1.62	66.42
641mm	nr	1.20	11.52	64.54	1.90	77.96
743mm	nr	1.25	12.00	75.04	2.18	89.22
844mm	nr	1.25	12.00	85.45	2.44	99.89
946mm	nr	1.30	12.48	95.95	2.71	111.14
1047mm	nr	1.35	12.96	103.22	2.90	119.08
1149mm	nr	1.35	12.96	113.63	3.16	129.75
1352mm	nr	1.40	13.44	133.93	3.68	151.05
1556mm	nr	1.45	13.92	154.53	4.21	172.66
1759mm	nr	1.50	14.40	174.93	4.73	194.06
1962mm	nr	1.55	14.88	195.74	5.27	215.89
2165mm	nr	1.60	15.36	216.34	5.79	237.49

ALTERATIONS AND REPAIRS

	Unit	Labour hours	Net labour (£)	Net material (£)	O'heads /profit (£)	Total (£)

NOTE

The prices contained within this section include the removal of all surplus materials from the site. No allowance has been made in the Labour hours column for any travelling time to and from the job or any disruption to the normal working day.

RAINWATER INSTALLATION

Remove existing gutters, joints and fittings, prepare to receive new gutterwork

	Unit	Labour hours	Net labour (£)	Net material (£)	O'heads /profit (£)	Total (£)
Cast iron						
up to 100mm	m	0.50	4.80	0.00	0.72	5.52
over 100mm	m	0.60	5.76	0.00	0.86	6.62
Asbestos cement						
up to 100mm	m	0.45	4.32	0.00	0.65	4.97
over 100mm	m	0.55	5.28	0.00	0.79	6.07

Remove 1.83m length of existing gutter, prepare ends and install new length of gutter to existing brackets

	Unit	Labour hours	Net labour (£)	Net material (£)	O'heads /profit (£)	Total (£)
Cast iron, half-round						
100mm	nr	2.00	19.20	12.53	0.79	32.52
115mm	nr	2.10	20.16	13.05	0.83	34.04
125mm	nr	2.15	20.64	15.30	0.90	36.84
Cast iron, ogee						
100mm	nr	2.00	19.20	13.69	0.82	33.71
115mm	nr	2.10	20.16	15.35	0.89	36.40
125mm	nr	2.15	20.64	16.10	0.92	37.66

	Unit	Labour hours	Net labour (£)	Net material (£)	O'heads /profit (£)	Total (£)
Aluminium, half-round						
100mm	nr	1.90	18.24	12.11	0.76	31.11
113mm	nr	2.00	19.20	12.83	0.80	32.83
125mm	nr	2.05	19.68	15.14	0.87	35.69
Aluminium, ogee						
100mm	nr	1.29	12.38	15.93	0.71	29.02
113mm	nr	2.00	19.20	18.46	0.94	38.60
125mm	nr	2.05	19.68	21.03	1.02	41.73
Remove existing gutter fittings, prepare ends and install new fitting						
Cast iron, half-round, 115mm						
angle	nr	1.00	9.60	5.24	0.37	15.21
outlet	nr	1.00	9.60	5.24	0.37	15.21
stop end	nr	0.50	4.80	2.24	0.18	7.22
Cast iron, ogee, 115mm						
angle	nr	1.00	9.60	4.84	0.36	14.80
outlet	nr	1.00	9.60	4.84	0.36	14.80
stop end	nr	0.50	4.80	2.06	0.17	7.03
Aluminium, half-round, 113mm						
angle	nr	0.95	9.12	4.52	0.34	13.98
outlet	nr	0.95	9.12	4.86	0.35	14.33
stop end	nr	0.45	4.32	1.96	0.16	6.44
Aluminium, ogee, 113mm						
angle	nr	0.95	9.12	5.44	0.36	14.92
stop end outlet	nr	0.95	9.12	4.38	0.34	13.84
stop end	nr	0.45	4.32	2.21	0.16	6.69

	Unit	Labour hours	Net labour (£)	Net material (£)	O'heads /profit (£)	Total (£)
Remove existing brackets from gutters and replace with galvanized steel repair brackets at 1m maximum centres						
Cast iron, half-round						
100mm	m	0.25	2.40	1.59	0.10	4.09
112mm	m	0.27	2.59	1.59	0.10	4.28
125mm	m	0.30	2.88	1.73	0.12	4.73
Cast iron, ogee						
100mm	m	0.25	2.40	1.70	0.10	4.20
112mm	m	0.27	2.59	1.70	0.11	4.40
125mm	m	0.30	2.88	1.94	0.12	4.94
Remove pipes, joints and fittings, prepare to receive new pipework						
Cast iron						
up to 100mm	m	0.50	4.80	0.00	0.72	5.52
over 100mm	m	0.60	5.76	0.00	0.86	6.62
Asbestos cement						
up to 100mm	m	0.45	4.32	0.00	0.65	4.97
over 100mm	m	0.55	5.28	0.00	0.79	6.07
Remove 1.83m length of existing pipe, replace with new						
Cast iron						
65mm	nr	1.80	17.28	24.65	1.05	42.98
75mm	nr	1.90	18.24	24.65	1.07	43.96
100mm	nr	2.05	19.68	33.06	1.32	54.06

	Unit	Labour hours	Net labour (£)	Net material (£)	O'heads /profit (£)	Total (£)
Remove existing pipe fittings, replace with new						
Cast iron shoes, eared						
65mm	nr	0.66	6.34	11.62	0.45	18.41
75mm	nr	0.80	7.68	11.62	0.48	19.78
100mm	nr	0.90	8.64	15.11	0.59	24.34
Cast iron bends						
65mm	nr	0.75	7.20	7.13	0.36	14.69
75mm	nr	0.95	9.12	8.24	0.43	17.79
100mm	nr	1.00	9.60	11.63	0.53	21.76
Rake out existing cement mortar joints to underground drain connection, re-form in cement mortar						
65mm	nr	0.40	3.84	0.57	0.11	4.52
75mm	nr	0.44	4.22	0.63	0.12	4.97
100mm	nr	0.50	4.80	0.68	0.14	5.62
HOT AND COLD WATER SUPPLY AND HEATING INSTALLATION						
Cut out 500mm length of copper pipe, Table Z, install new length of pipe with new brass compression connections to existing ends						
Copper						
15mm	nr	0.67	6.43	3.28	0.24	9.95
22mm	nr	0.69	6.62	5.21	0.30	12.13
28mm	nr	0.83	7.97	9.47	0.44	17.88
35mm	nr	0.96	9.22	18.48	0.69	28.39
42mm	nr	1.01	9.70	23.56	0.83	34.09
54mm	nr	1.12	10.75	34.46	1.13	46.34

	Unit	Labour hours	Net labour (£)	Net material (£)	O'heads /profit (£)	Total (£)
Remove existing pipework, prepare to receive new pipework						
Copper						
up to 25mm	m	0.35	3.36	0.00	0.50	3.86
25-54mm	m	0.50	4.80	0.00	0.72	5.52
Lead						
up to 1in	m	0.35	3.36	0.00	0.50	3.86
Take off existing radiator valve, replace with new, drain down system prior to removal and bleed after installation						
Single valves, standard type	nr	1.50	14.40	5.01	0.49	19.90
Complete systems, 9nr valves, standard type	nr	4.60	44.16	45.08	2.23	91.47
Take out existing galvanized steel water storage tank, install new plastic tank complete with ball valve, lid and insulation, allow for cutting holes, tank connectors, make-up pipework and connectors to existing pipework						
18 litres tank	nr	4.00	38.40	35.13	1.84	75.37
68 litres tank	nr	4.50	43.20	47.64	2.27	93.11
154 litres tank	nr	5.50	52.80	79.92	3.32	136.04

Take off and release existing radiator, drain down system prior to removal and bleed after installation, allow for new surface-mounted pipework and removing and refixing existing brackets, compression connections to existing pipework

	Unit	Labour hours	Net labour (£)	Net material (£)	O'heads /profit (£)	Total (£)
Distance of relocation						
not exceeding 2m	nr	2.50	24.00	9.56	0.84	34.40
2-3m	nr	2.75	26.40	11.36	0.94	38.70
3-4m	nr	3.00	28.80	12.72	1.04	42.56
4-5m	nr	3.25	31.20	15.96	1.18	48.34
SANITARY FITTINGS						
Take out cast iron bath and trap, cut back pipework, prepare to receive new, install new plastic bath with trap, integral overflow, pair of pillar taps, waste, plug, make up pipework and connect to existing hot and cold water and waste pipework						
new bath, 1700mm long	nr	8.00	76.80	190.32	40.07	307.19
Take out existing wash basin including trap, cut back pipework and prepare to receive new, install new pedestal-mounted wash basin with trap, pair of pillar taps, pipework as necessary and connections to hot and cold water and waste pipework						
new wash basin, size 590 x 475mm	nr	3.75	36.00	111.30	22.10	169.40
Take out existing high level WC and connecting pipework including overflow, cut off spigot, new low level WC suite complete including new overflow through backwall, Quikfit connector to soil pipe and connection to cold water supply						
new WC suite, size 780 x 510 x 710mm	nr	4.50	43.20	175.86	32.86	251.92

	Unit	Labour hours	Net labour (£)	Net material (£)	O'heads /profit (£)	Total (£)
WORK TO EXISTING FITTINGS						
Remove existing tap and valves and replace with new acrylic head conversion kit	nr	0.30	2.88	13.65	0.41	16.94
Remove existing tap heads and valves and replace tap washers	nr	0.30	2.88	1.10	0.10	4.08
Remove existing pillar taps to bathroom sink and bath, replace with new	nr	1.20	11.52	44.08	1.39	56.99
Remove and replace existing plastic diaphragm to washdown cistern	nr	0.50	4.80	1.16	0.15	6.11

APPROXIMATE ESTIMATING

APPROXIMATE ESTIMATING

The prices in this section are intended as a guide to the approximate estimation of costs for total elements of installation. The prices are inclusive of 15% overheads and profit.

RAINWATER INSTALLATION

uPVC, 112mm half-round gutter with 75mm downpipe

	£
Terrace house 7m wide with two gutters, outlets, downpipes, offsets and shoes	200
Semi-detached house with hip end, three gutters, outlets, downpipes, offsets and shoes	295
Semi-detached house with gable end, two gutters, outlets, downpipes, offsets and shoes	210
Detached house with gable ends, two gutters, outlets, downpipes, offsets and shoes	220
Detached house with hip ends, all round four gutters, outlets, downpipes, offsets and shoes	290

uPVC, 112mm half-round gutter with 75mm downpipe (cont'd)	£
Bungalow with gable ends, two gutters, outlets, downpipes, offsets and shoes	195
Cast-iron 115mm half-round gutter with 75mm downpipes with ears	
Terrace house, 7m wide with two gutters, outlets, downpipes, offsets and shoes	475
Semi-detached house 8m x 10m with hip end, three gutters, outlets, downpipes, offsets and shoes	610
Semi-detached house 8m x 10m with gable end, two gutters, outlets, downpipes, offsets and shoes	495
Detached house 9m x 11m with gable ends, two gutters, outlets, downpipes, offsets and shoes	735
Detached house 9m x 11m with hip ends all round, four gutters, outlets, downpipes, offsets and shoes	635
Bungalow 10m x 10m with gable ends, two gutters, outlets, downpipes, offsets and shoes	580
Aluminium 113mm half-round gutter: 75mm downpipes	
Terrace house 7m wide with two gutters, outlets, downpipes, offsets and shoes	440

Aluminium 113mm half-round gutter: 75mm downpipes (cont'd)

Semi-detached house 8m x 10m with hip end, three gutters, outlets, downpipes, offsets and shoes	575
Semi-detached house 8m x 10m with gable end, two gutters, outlets, downpipes, offsets and shoes	460
Detached house 9m x 11m with gable ends, two gutters, outlets, downpipes, offsets and shoes	485
Detached house 9m x 11m with hip ends all round, four gutters, outlets, downpipes, offsets and shoes	630
Bungalow 10m x 10m with gable ends, two gutters, outlets, downpipes, offsets and shoes	535

HOT AND COLD WATER INSTALLATION

Hot and cold water installation complete; plastic storage cistern; rising main; down service; gas multi-point hot water heater supplies to bathroom and kitchen, sanitary fittings, sink unit and wastes, copper pipework

Semi-detached house (no central heating)	1,450

Hot and cold water installation complete; gas-fired boiler and pump, with plastic storage cistern and header tank; seven radiators, thermostat, time clock, rising main, down service, indirect copper cylinder with immersion heater; hot water supplies to bathroom and kitchen; sanitary fittings, sink unit and wastes, copper pipework

Semi-detached house	2,900
Add for additional cloakroom on ground floor with close-coupled WC and wash basin	400
Add for four bedroom detached house with en-suite master bedroom upgraded central heating system, coloured siphonic sanitary fittings; shower unit and bidet	2,775

SOIL AND WASTE INSTALLATION

Soil and waste installation complete; serving first floor bathroom with WC, bath, wash basin

uPVC pipework	250
Cast iron pipework with caulked lead joints	500
Cast iron pipework with 'Timesaver' joints	430

SANITARY APPLIANCES

Wash basin 560 x 430mm in vitreous china; chromium-plated pillar taps, waste plug and chain on brackets	130
Extra for pedestal	55
Extra for larger coloured basin 680 x 530mm with pedestal	300

Bath complete with chromium-plated pillar taps, waste, plug and chain, overflow and trap, bath panel to one side and one end

Acrylic white 1700 x 700mm	260
Coloured	290

Steel porcelain enamelled 1700 x 700mm white	260
Heavy-duty steel porcelain enamelled 1700 x 700mm white	440
Extra for coloured finish	50
Whirlpool bath complete	2,000
Washdown WC suite complete with close-coupled cistern, seat and overflow pipe, white vitreous china	210
Extra for siphonic type	140
Luxury coloured siphonic WC suite complete	390
Asian squatting pan with high-level flushing cistern complete with flexible hose washing pipe	440
White vitreous china bidet, with monobloc mixer tap	285
Extra for coloured	50
White vitreous china urinal bowl with exposed pipes chromium domed grating (no cistern)	180
Extra for coloured	35
Fireclay cleaners sink 450 x 400mm, complete with grating and pad, taps, legs and supports, grid waste plug and chain	340
Fireclay 'Belfast' sink 535 x 405 x 255mm, complete with taps (no supports)	230

Chapter 6
Plant and tool hire

The prices contained in this chapter are based upon information supplied by HSS Hire Shops Ltd, 25 Willow Lane, Mitcham, Surrey who have over 160 shops throughout the country (see Yellow Pages). The prices exclude VAT and delivery charges.

	First 24 hrs £	Addit 24 hrs £	Per week £
ACCESS AND SUPPORT			
Narrow tower base, size 1.3 x 1.5m, height			
2.5m	27.50	13.75	55.00
4.5m	40.50	20.25	81.00
6.5m	53.50	26.75	107.00
8.5m	66.50	33.25	133.00
10.5m	79.50	39.75	159.00
Span tower base, size 1.3 x 1.5m or 1.3 x 2.5m, height			
2.5m	27.50	13.75	53.00
4.5m	40.50	20.25	81.00
6.5m	53.50	26.75	107.00
8.5m	66.50	33.25	133.00
10.5m	79.50	39.75	159.00

Access and support (cont'd)	First 24 hrs	Addit 24 hrs	Per week
Alloy ladders			
Double 3.5m extending to 6.2m	9.00	4.50	18.00
5.0m 9.0m	12.50	6.25	25.00
Treble 2.5m 6.0m	9.00	4.50	18.00
3.5m 9.1m	12.50	6.25	25.00
Roof ladders			
Alloy 4.9m, 5.9m and 6.9m	15.00	5.00	25.00
Ladder stay each	5.40	1.80	9.00
Builder's steps			
8 Tread, height 1.5m	7.00	3.50	14.00
10 Tread, height 2.1m	8.00	4.00	16.00
12 Tread, height 2.7m	9.00	4.50	18.00
Steel trestles			
Nos 1-4: 0.5m extending to 2.4m	-	-	3.20
Scaffold boards			
Length 2.4-3.9m	-	-	1.90
Lightweight staging			
Length : 2.4m	8.00	4.00	10.00
3.0m	9.00	4.50	11.25
3.6m	10.00	5.00	12.50
4.2m	11.00	5.50	13.75
4.8m	13.00	6.50	16.25
6.0m	15.00	7.50	18.75
7.2m	20.00	10.00	25.00

BUILDING AND DECORATING	First 24 hrs £	Addit 24 hrs £	Per week £
Metal locator	9.00	3.00	15.00
Cable avoiding tool	32.40	10.81	54.00
Cat signal generator	19.20	6.40	32.00
Steel props Nos 0-4: 1.8m extending to 4.9m (quantity discounts)	-	-	3.20
Jackall prop	9.60	3.20	16.00
Building tools			
Tile saw	38.40	11.60	58.00
Hand operated tile cutter	8.40	2.80	14.00
Blowlamp with extension hose	6.60	2.20	11.00
Bolt croppers	7.20	2.40	12.00
Crowbar	3.60	1.20	6.00
Floorboard cramps	4.20	1.40	7.00
G-cramps, sash cramps	3.00	1.00	5.00
Pickaxe matlock, punner	3.60	1.20	6.00
Shovel/spade	3.60	1.20	6.00
Site tool box	15.60	5.20	26.00
Sledgehammer 3kg and 6.5kg	3.60	1.20	6.00
Spirit level	3.60	1.20	6.00
Tarpaulins	7.20	2.40	12.00
Wheelbarrow	3.60	1.20	6.00
Workmate	9.00	3.00	15.00
Road hazard equipment			
Traffic warning cones (0.5m)	-	-	1.60
Flashing lamp and stand	-	-	5.00
Road signs	-	-	9.50
Road barrier, cone and cap	-	-	8.00

	First 24 hrs £	Addit 24 hrs £	Per week £
DRAIN CLEARING/PLUMBING/PUMPING			
Plumber's tools			
Blowtorch	7.20	2.40	12.00
Pipe freezing kit CO2 (gas)	21.00	7.00	24.50
Steel pipe bender, hydraulic	31.20	10.40	52.00
Copper pipe bender			
hand held	9.00	3.00	15.00
large	19.20	6.40	32.00
Steel pipe cutter	6.60	2.20	11.00
Clay pipe cutter	12.00	4.60	20.00
Pipe vice bench	9.60	3.20	16.00
Pipe wrenches			
450mm	4.80	1.60	8.00
600mm	6.60	2.20	11.00
900mm	7.80	2.60	13.00
Chain wrench	9.60	3.20	16.00
Pipe threading			
Electric die stock	39.00	45.50	65.00
Ratchet die stock	15.60	5.20	26.00
Pipe-threading machine ½-4in electric	72.00	24.00	120.00
Pipe-threading machine ½-2in electric	54.00	18.00	90.00
Pipe pressure tester	15.60	5.20	26.00
Pipe saw	21.60	7.20	36.00
Water pumps			
Submersible 25mm, electric	12.00	4.00	20.00
Submersible 32mm, electric	18.00	6.00	30.00
Submersible 50mm, petrol	27.60	9.20	46.00
Centrifugal 50mm, petrol	31.20	10.40	52.00
Centrifugal 75mm, petrol	39.00	13.00	65.00
Drain test kit, 'U' gauge	6.00	2.00	10.00
Air bag drain stopper	7.20	2.40	12.00

Drain testing/clearing tools	First 24 hrs £	Addit 24 hrs £	Per week £
Drain plugs 100-150mm (per pair)	3.60	1.20	6.00
Drain rods and fittings 9m (per set)	8.40	2.80	14.00
Ropump	8.40	2.80	14.00
Sink cleaner, hand-operated	8.40	2.80	14.00
Drain cleaner, hand-operated	15.00	5.00	25.00
Powered drain cleaner, electric	34.80	11.60	58.00

HEATING, COOLING AND DRYING

Industrial heaters - gas			
Plaque heater output 2,500 Btu	10.80	3.60	18.00
Forced air output 140,000 Btu	28.80	9.60	48.00
Forced air output 275,000 Btu	39.00	13.00	65.00
Industrial heaters - paraffin			
Forced air, 60,000 Btu	27.00	9.00	45.00
Forced air, 85-100,000 Btu	30.00	10.00	50.00
Forced air, 150,000 Btu	39.00	13.00	65.00
Home/office heaters			
Cabinet, gas, 3-16,000 Btu	8.40	2.80	14.00
Electric fan heater - 3kW	6.00	2.00	10.00
Air conditioning units			
Air conditioner	57.00	19.00	95.00
Space cooler	51.00	17.00	85.00
Cold air blower	21.00	7.00	35.00
Drying			
Building dryer dehumidifier - 300m3	39.00	13.00	65.00
Portable building dryer - 140m3	22.80	7.60	38.00
Portable fume extractor	42.00	14.00	70.00

	First 24 hrs £	Addit 24 hrs £	Per week £
LIGHTING, WELDING AND POWER			
Lighting units			
Gas - large tripod-mounted	12.00	4.00	20.00
Gas - small cylinder-mounted	9.00	3.00	15.00
Electric - tripod mounted	12.00	4.00	20.00
Plasterers' light - fluorescent	12.00	4.00	20.00
Magnetic 500W flood	11.40	3.80	19.00
Twin 500W flood - 5m tower mast	21.60	7.20	36.00
Festoon lights, industrial (34cm)	12.00	4.00	20.00
Welding			
Arc welder 140/180 amp, 240 volt	19.20	6.40	32.00
MIG welder, 240 volt, medium	25.20	8.40	29.40
No-gas MIG welder	19.20	6.40	32.00
Spot welder, 240 volt (30 amp)	19.20	6.40	32.00
Site welder 20-170 amp, petrol	46.80	15.60	78.00
Welder/generator 300 amp, d.c. silenced	96.00	32.00	160.00
Oxy/acetylene welding kit	28.80	9.60	48.00
Generators (continuous rating)			
3 kVA petrol, 110/240 volt	34.80	11.60	58.00
4 kVA diesel, 110/240 volt	46.80	15.60	78.00
8 kVA diesel, 110/240 volt	72.00	24.00	120.00
15 kVA diesel, 110/240 volt	90.00	30.00	150.00
Transformers			
2.2 kVA	5.70	1.90	9.50
3.0 kVA	8.40	2.80	14.00
5.0 kVA	15.60	5.20	26.00
10 kVA	29.40	9.80	49.00

	First 24 hrs	Addit 24 hrs	Per week
Extension cable, 15m cable/drum	4.50	1.50	7.50
Fourway junction box	7.20	2.40	12.00
Power breaker RCD plug	3.00	1.00	5.00
KANGO BREAKING AND DRILLING			
Hydraulic breakers			
Heavy duty, diesel	57.00	19.00	95.00
Medium duty, petrol	48.00	16.00	80.00
Electric hammers			
Heavy breaker (inc. trolley)	33.60	11.20	56.00
Medium duty breaker	18.00	6.00	30.00
Rotary hammers			
Hilti breaker drill, TE 72	18.00	6.00	30.00
Hilti breaker drill, TE 52	18.00	6.00	30.00
Medium duty breaker drill	16.20	5.40	27.00
Light duty	16.20	5.40	27.00
Electric drills			
Two speed percussion drill	8.40	2.80	14.00
Cordless drill	12.00	4.00	20.00
Right angle drill	16.20	5.40	27.00
Four speed drill	16.20	5.40	27.00
Magnetic base drills			
Magnetic drill stand (drill extra)	30.60	10.20	51.00
Magnetic drill stand c/w drill	48.80	15.60	50.60

	First 24 hrs	Addit 24 hrs	Per week
FIXING, GRINDING AND SANDING			
Fixing tools			
Cartridge hammer	18.00	6.00	30.00
Staple tacker - light duty	6.00	2.00	10.00
Hammer stapler - heavy duty	13.20	4.40	22.00
Nail gun - air operated, automatic	24.00	8.00	40.00
Impact wrench - electric	15.20	5.20	26.00
Screwdriver - electric, auto-feed	16.80	5.60	28.00
Angle grinders			
Angle grinder, 100mm	8.40	2.80	14.00
Angle grinder, 230mm	11.40	3.80	19.00
Angle grinder, 300mm	20.40	6.80	34.00
Sanders			
Belt sanders	15.00	5.00	25.00
Disc sander, 180mm	15.00	5.00	25.00
Orbital sander industrial	14.40	4.80	24.00
Floor sanders			
Domestic, 200mm	24.60	8.20	41.00
Edging sander	21.60	7.20	36.00
Floor sander c/w edging sander	37.50	12.50	62.50
Router			
Laminate trimmer	16.80	5.60	28.00
Power plane	16.80	5.60	28.00

	First 24 hrs	Addit 24 hrs	Per week
SAWING AND CUTTING			
Timber saw benches			
Combination bench, 300mm	51.00	17.00	85.00
Combination bench, 200mm	39.00	13.00	65.00
General-purpose saws			
Crosscut and mitre saw	28.80	9.60	48.00
Flipover mitre saw	39.00	13.00	65.00
Band saw	49.20	16.40	82.00
Reciprocating saw	17.40	5.80	29.00
Jig saw	14.40	4.80	24.00
Circular saw, 230mm blade	15.60	5.20	26.00
Circular saw, 150mm blade	12.60	4.20	21.00
Chain saw with safety kit	39.00	15.00	65.00
Metal and masonry cutting saw			
Masonry sawbench, petrol	54.00	18.00	90.00
Masonry saw bench, electric	45.00	15.00	75.00
Bench top cut-off saw	28.80	9.60	48.00
Portable cut-off saw, electric/petrol	20.40	6.80	34.00
Cut-off saw trolley	6.00	2.00	10.00
Chasing machine	36.00	12.00	60.00
Metal shears	19.20	6.40	32.00
Metal nibblers	19.20	6.40	32.00
Floor saw, 450mm	51.60	17.20	86.00
Floor saw, 350mm	39.00	13.00	65.00
Tile saw	34.80	11.60	58.00
Slab splitter	32.40	10.80	54.00
Block splitter	16.80	5.60	28.00

	First 24 hrs	Addit 24 hrs	Per week
LIFTING AND MATERIALS HANDLING			
Man lift, height 7.3m	75.00	25.00	125.00
Man lift, height 9.1m	120.00	40.00	200.00
Tirfor winch, TU 16	15.60	5.20	26.00
Tirfor winch, TU 32	19.80	6.60	33.00
Scaffold hoist (200kg)	45.00	15.00	75.00

Chapter 7
General data

The metric system

Linear

1 centimetre (cm)	=	10 millimetres (mm)
1 decimetre (dm)	=	10 centimetres (cm)
1 metre (m)	=	10 decimetres (dm)
1 kilometre (km)	=	1000 metres (m)

Area

100 sq millimetres	=	1 sq centimetre
100 sq centimetres	=	1 sq decimetre
100 sq decimetres	=	1 sq metre
1000 sq metres	=	1 hectare

Capacity

1 millilitre (ml)	=	1 cubic centimetre (cm3)
1 centilitre (cl)	=	10 millilitres (ml)
1 decilitre (dl)	=	10 centilitres (cl)
1 litre (l)	=	10 decilitres (dl)

Weight

1 centigram (cg)	=	10 milligrams (mg)
1 decigram (dg)	=	10 centigrams (cg)
1 gram (g)	=	10 decigrams (dg)
1 decagram (dag)	=	10 grams (g)
1 hectogram (hg)	=	10 decagrams (dag)
1 kilogram (kg)	=	10 hectograms (hg) = 1000 grams (g)

GENERAL DATA

Imperial/metric conversions

Linear

1 in = 25.4mm	1mm = 0.03937in
1 ft = 304.8mm	1cm = 0.3937in
1 yd = 914.4mm	1dm = 3.937in
	1m = 39.37in

Square

1 sq in = $645.16mm^2$	$1cm^2$ = 0.155 sq in
1 sq ft = $0.0929m^2$	$1m^2$ = 10.7639 sq ft
1 sq yd = $0.8361m^2$	$1m^2$ = 1.196 sq yd

Cube

1 cu in = $16.3871cm^3$	$1cm^3$ = 0.061 cu in
1 cu ft = $0.0283m^3$	$1m^3$ = 35.3148 cu ft
1 cu yd = $0.7646m^3$	$1m^3$ = 1.307954 cu yd

Capacity

1 fl oz = 28.4ml	1ml = 0.0352 fl oz
1 pt = 0.568 ltr	1dl = 3.52 fl oz
1 gallon = 4.546 ltr	1 ltr = 1.7598 pt

Weight

1 oz = 28.35g	1g = 0.035 oz
1 lb = 0.4536kg	1kg = 35.274 oz
1 st = 6.35kg	1t = 2204.6 lb
1 ton = 1.016t	1t = 0.9842 ton

Temperature equivalents

In order to convert Fahrenheit to Celsius deduct 32 and multiply by 5/9. To convert Celsius to Fahrenheit multiply by 9/5 and add 32.

Fahrenheit	Celsius
230	110.0
220	104.4

Fahrenheit	Celsius
210	98.9
200	93.3
190	87.8
180	82.2
170	76.7
160	71.1
150	65.6
140	60.0
130	54.4
120	48.9
110	43.3
90	32.2
80	26.7
70	21.1
60	15.6
50	10.0
40	4.4
30	-1.1
20	-6.7
10	-12.2
0	-17.8

Melting points of materials

Aluminium	658°C
Brass	927-1010°C
Bronze	912°C
Cast iron	1186°C
Copper	1083°C
Lead	327°C
Nickel	1452°C
Steel	1371°C
Tin	230°C
Zinc	419°C

Standard wire gauge and metric equivalent

SWG	mm	SWG	mm
3	6.40	15	1.83
4	5.89	16	1.63
5	5.38	17	1.42
6	4.88	18	1.21
7	4.47	19	1.02
8	4.06	20	0.91
9	3.63	21	0.81
10	3.25	22	0.71
11	2.95	23	0.61
12	2.65	24	0.56
13	2.34	25	0.51
14	2.03	26	0.46

LEADWORK

Range of thicknesses and weights to BS1178

Code	Thickness mm	Weight kg/m2	Code	Thickness mm	Weight kg/m2
Green (3)	1.32	14.97	Black (6)	2.65	30.05
Blue (4)	1.80	20.41	White (7)	3.15	36.72
Red (5)	2.24	25.40	Orange (8)	3.55	40.26

Thickness of lead sheet for various purposes

Use	Code
Soakers, apron and cover flashings, hip and ridge flashings, pitched valley gutters	Green or Blue
Weatherings to cornices, parapets etc., lead slates, chimney flashings - back gutters, apron and side flashings	Blue or Red
Parapet, box and tapered gutters	Red to Orange

Sizing. Guidelines to good practice

Recommended maximum sizes of individual pieces of lead for the following applications.

a. Flashings - apron and cover. Also pitched valley gutter linings

Code	Uses	Maximum length mm
Green (3)	Soakers	1000
Blue and Red (4 or 5)	Flashings and pitched valley gutters	1500

b. Gutters - box, tapered and parapet

Code		Maximum length between drips mm	Overall girth Max. mm
Blue	(4)	1500	750
Red	(5)	2000	800
Black	(6)	2250	850
White	(7)	2500	900
Orange	(8)	3000	1000

Zinc sheets to BS849

Zinc No. gauge	2134 x 914mm Weight per sheet kg	2438 x 914mm Weight per sheet kg
10	5.8	7.7
11	7.9	9.0
12	8.9	10.2
13	10.0	11.4
14	11.0	12.7
15	12.8	14.6

Nos 14 to 16 for roofing. Nos 15 to 16 for gutters

COPPER TUBES, SIZE AND PRESSURE

BS2871: Part 1: 1971

	TABLE X Based on material in half hard condition		TABLE Z Based on material in hard condition	
Size of tube mm	Nominal thickness mm	Maximum working pressures bar	Nominal thickness mm	Maximu working pressure bar
6	0.6	133	0.5	133
8	0.6	97	0.5	10
10	0.6	77	0.5	78
12	0.6	63	0.5	64
15	0.7	58	0.5	50
18	0.8	56	0.6	50
22	0.9	51	0.6	41
28	0.9	40	0.6	32
35	1.2	42	0.7	30
42	1.2	35	0.8	28
54	1.2	27	0.9	25
67	1.2	20	1.0	20
76.1	1.5	24	1.2	19
108	1.5	17	1.2	17
133	1.5	14	1.5	16
159	2.0	15	1.5	15

S2871: Part 1: 1971

	TABLE Y		TABLE W		
		Half hard	Annealed	Annealed	
ize of tube	Nominal thickness	Maximum working pressures	Maximum working pressures	Preferred thickness	Maximum working pressures
mm	mm	bar	bar	mm	bar
6	0.8	188	144	0.6	90
8	0.8	136	103	0.6	66
10	0.8	106	82	0.7	62
12	0.8	87	67	-	-
15	1.0	87	67		
18	1.0	72	55		
22	1.2	69	57		
28	1.2	55	42		
35	1.5	54	41		
42	1.5	45	34		
54	2.0	47	36		
67	2.0	37	28		
76.1	2.0	33	25		
108	2.5	29	22		

PVC PRESSURE PIPE SYSTEMS

iameter and wall thickness dimensions

ipe for cold water services to BS3505: 1986, standard pipe length: 6m effective ngth.

Nominal size	Mean outside diameter Min. mm	Mean outside diameter Max. mm	Individual outside diameter Min. mm	Individual outside diameter Max. mm	Wall thickness 12 bar (Class D) Average value Max. mm	Wall thickness 15 bar (Class E) Average value Max. mm
3/8in	17.0	17.3	17.0	17.3	-	1.9
1/2in	21.2	21.5	21.2	21.5	-	2.1
3/4in	26.6	26.9	26.6	26.9	-	2.5
1in	33.4	33.7	33.3	33.8	-	2.7
1 1/4in	42.1	42.4	42.0	42.5	2.7	3.2
1 1/2in	48.1	48.4	48.0	48.5	3.0	3.7

PIPE FOR POTABLE WATER SUPPLY AND DISTRIBUTION TO BS3505: 1986

Nominal size	Mean outside diameter Min. mm	Mean outside diameter Max. mm	Individual outside diameter Min. mm	Individual outside diameter Max. mm	Wall thickness 9 bar (Class C) Average value Max. mm	Wall thickness 12 bar (Class D) Average value Max. mm
2in	60.2	60.5	60.0	60.7	3.0	3.7
3in	88.7	89.1	88.4	89.4	4.1	5.3
4in	114.1	114.1	113.7	114.9	5.2	6.8
5in	140.0	140.4	139.4	141.0	6.3	8.3
6in	168.0	168.5	167.4	169.1	7.5	9.9
8in	218.8	219.4	218.0	220.2	8.8	11.6
9in	244.1	244.8	243.1	245.8	9.8	12.9
10in	272.6	273.4	271.6	274.4	10.9	14.3
12in	323.4	324.3	322.2	325.5	12.9	17.0
14in	355.0	356.0			13.7	18.0
16in	405.9	406.9	404.3	408.5	16.2	21.1
18in	456.7	457.7			17.6	23.1

uPVC RAINWATER SYSTEMS

Sizes and weights

Gutters	Size	Wall thickness	Weight per metre
75mm (3in) HR	75mm	2.15mm	0.34kg
100mm (4in) HR	101.6mm	2.3mm	0.53kg
125mm (5in) HR	125mm	2.3mm	0.68kg
150mm (6in) HR	165mm	2.6mm	0.96kg
112mm (4in) RR	112mm	2.16mm	0.62kg

Downpipes	Size	Wall thickness	Weight per metre
50mm (2in) dia	50mm	1.6mm	0.36kg
63mm (2in) dia	63mm	1.9mm	0.54kg
75mm (3in) dia	75mm	2.0mm	0.60kg
100mm (4in) dia	110mm	3.175mm	1.70kg
65mm (2in) sq	65mm	2.16mm	0.63kg

INDEX